CONTROL VALVE PRIMER

A USER'S GUIDE

FOURTH EDITION

Dear John, I hope that
this book may be of some use
With best regards,

CONTROL VALVE PRIMER

A USER'S GUIDE

FOURTH EDITION

HANS D. BAUMANN

12/17/2011

Notice

The information presented in this publication is for the general education of the reader. Because neither the author nor the publisher have any control over the use of the information by the reader, both the author and the publisher disclaim any and all liability of any kind arising out of such use. The reader is expected to exercise sound professional judgment in using any of the information presented in a particular application.

Additionally, neither the author nor the publisher have investigated or considered the effect of any patents on the ability of the reader to use any of the information in a particular application. The reader is responsible for reviewing any possible patents that may affect any particular use of the information presented.

Any references to commercial products in the work are cited as examples only. Neither the author nor the publisher endorses any referenced commercial product. Any trademarks or tradenames referenced belong to the respective owner of the mark or name. Neither the author nor the publisher makes any representation regarding the availability of any referenced commercial product at any time. The manufacturer's instructions on use of any commercial product must be followed at all times, even if in conflict with the information in this publication.

Printed in the United States of America.
10 9 8 7 6 5 4 3 2

ISBN-13: 978-1-934394-50-2

ISA
67 Alexander Drive
P.O. Box 12277
Research Triangle Park, NC 27709

Library of Congress Cataloging-in-Publication Data

Baumann, Hans D., 1930-
Control valve primer : a user's guide / Hans D. Baumann. -- 4th ed.
p. cm.
Includes bibliographical references.
ISBN 978-1-934394-50-2 (pbk.)
1. Pneumatic control valves--Handbooks, manuals, etc. I. Title.
TJ223.V3B38 2008
629.8'045--dc22

2008024257

DEDICATION

This book is dedicated to my wife, Sigrid, for her lifetime of support and to the many friends at ISA for their help and encouragement I received throughout more than forty years.

TABLE OF CONTENTS

INTRODUCTION TO THE FOURTH EDITION

Keeping track of the ever evolving changes in industry, the fourth edition of the *Control Valve Primer* has been updated and significantly enlarged by adding the following new or revised materials:

- enlarged chapter on digital positioners,
- increased coverage of DCS Systems and smart positioners,
- a new hydrodynamic and cavitation noise estimate,
- updated standards to reflect latest IEC versions,
- a new chapter on seat leakage and seat materials,
- additional illustrations,
- updated standards listing, and
- updated Aerodynamic Noise equations.

The Editors

FOREWORD

This book is intended primarily for those novice instrument engineers who, even though they have been crammed full of feedback theory from their college control courses and consider themselves experts in computer programing, are suddenly faced with the challenge of designing a control loop in a process plant. Here, they discover that, in addition to all the vaguely familiar electronic hardware such as controllers and transmitters, they also need a final control element. This turns out to be a cross between a piece of plumbing hardware and a pneumatic logic system. Such people are usually ill-equipped to answer the 30-plus entry questions on a typical control valve specification sheet in the allotted time of, say, 10 minutes. Therefore, having a reference book at their disposal might make their job easier, and it might even help them to distinguish what is relevant when bombarded with conflicting claims from competing control valve vendors. At the very least, it may keep them from wasting valuable time by not having to wade through many hundreds of pages of scientifically worded handbook trivia.

While it is true that any personal computer can handle complex sizing equations, I still feel it is useful to have some simple formulas that can be handled by a pocket calculator if the need arises. This led me to propose short cuts in the sizing and noise equations.

The reader is hereby cautioned that the opinions expressed herein are strictly those of the writer, and no guarantee is given, nor implied, as to the accuracy of the book's contents. As a matter of fact, some of the statements are considered downright controversial. I am seldom accused of being unbiased.

It all started when, as president of one of the smaller control valve companies, I visited a major tobacco company in Richmond, Virginia in order to do some sales promotion. However, after having carried about half a dozen valve samples up to the second floor conference room, the leader of the group of young instrument engineers stated flatly, "Please no sales talk. Instead, answer some of my questions." As it turned out, he liked the answers, and at the end of the session suggested I write a book. Well, here it is.

ABOUT THE AUTHOR

Hans D. Baumann received an industrial engineering education in his native Germany and studied under U.S. Government sponsorship at Western Reserve University and later at Northeastern University, culminating in a Ph.D. in Mechanical Engineering from Columbia Pacific University. In addition, he is registered as a Professional Engineer in four states. During his professional career, he personally designed or directed the development of over 28 valve lines. One of them, the famous "CAMFLEX" valve and its derivations, is produced in eight countries where over three million units have been sold. He is credited with over 150 U.S. and worldwide patents and has published 115 papers and articles in addition to co-authoring seven handbooks on valves, instrumentation, and acoustic. He also was named by *InTech* magazine one of fifty most important innovators and wrote the acclaimed business book *The Ideal Enterprise*.

Prior to founding his own control valve manufacturing company in 1977, he was an International Consultant, Corporate Vice-President of Masoneilan-International, Inc., and Manager of R&D at Worthington S/A in France. After selling his company to Emerson Electric, he worked for Fisher Controls as Senior Vice President.

Usually ahead of his time, his "critical flow factor" and "pipe reducer correction factors" (F_L & F_p) for valve sizing, introduced in the early sixties, later became part of ISA sizing standards in 1972, and his proposal in 1970 to utilize modified jet noise theories for aerodynamic valve noise prediction, became the basis for the ISA-75.17-1989 and IEC standard 60534-8-3. He served as a director of the ISA Standards & Practices Department Board, Chairman of the ISA75.11 Committee, U.S. Technical Expert for IEC Committee SC/65B/WG9, was a Member of the ASME Bioprocessing Equipment Executive Committee, and Chairman of the Equipment Subcommittee on Seals, and was the former Standards Chairman for Control Valves for the Fluid Controls Institute.

As a well received guest speaker in the U.S. and abroad, he has also been invited as a guest Professor to Kobe University in Japan and to the Korean Advanced Institute of Technology in Korea.

As a Life Fellow Member of ISA and ASME, he has been honored for his "many contributions to the science and technology of control valves" with the "Chet Beard" and the "UOP Technology" awards; he was named Honorary Member of the Spanish Chemical Engineering Society, and Honorary Life Member of the Fluid Controls Institute. His many valve designs have been honored with a gold medal from Germany, prizes from France and Japan, and seven U.S. "Vaaler" awards. He is a member of Sigma Xi, the scientific research society.

1

WHAT IS A CONTROL VALVE AND HOW DOES IT AFFECT MY CONTROL LOOP?

Control valves may be the most important, but sometimes the most neglected, part of a control loop. The reason is usually the instrument engineer's unfamiliarity with the many facets, terminologies, and areas of engineering disciplines such as fluid mechanics, metallurgy, noise control, and piping and vessel design that can be involved depending on the severity of service conditions.

Any control loop usually consists of a sensor of the process condition, a transmitter, and a controller that compares the "process variable" received from the transmitter with the "set point," i.e., the desired process condition. The controller, in turn, sends a corrective signal to the "final control element," the last part of the loop and the "muscle" of the process control system. While the "sensors" of the process variables are the eyes, the "controller" the brain, then the "final control element" is the hands of the control loop. This makes it the most important, alas sometimes the least understood, part of an automatic control system. This comes about, in part, due to our strong attachment to electronic systems and computers causing some neglect in the proper understanding and proper use of the all important hardware.

Control valves are the most common type of final control elements; however, there are other types such as:

- devices that regulate (throttle) electric energy such as silicon-controlled rectifiers,
- variable speed drives,
- feeders, pumps, and belt drives, and
- dampers.

Some of these devices perform functions similar to control valves and could be used as an alternative. For example, in order to control the pH level, a variable stroke-type metering pump may be used to inject acid into wastewater (instead of using a control valve lined with polytetrafluorethylene [PTFE]).

What then is a control valve? This is a difficult question since there is considerable overlap with other types of valves. For example, a valve operating strictly in the on-off mode (such as the hydronic solenoid valve in your home heating system) could be replaced by a simple ball valve operated by a pneumatic cylinder, a type usually referred to as an "automated valve."

The distinction between "automated" and "control" valves is usually considered to be the ability of the latter to "modulate," i.e., to assume an infinite number of "throttling" travel positions during normal control service.

Physically, there are three basic components of a control valve:

- The valve body subassembly. This is the working part and, in itself, a pressure vessel.
- The actuator. This is the device that positions the throttling element inside the valve body.
- Accessories. These are positioners, I/P transducers, limit switches, handwheels, air sets, position sensors, solenoid valves, and travel stops.

A more detailed breakdown of the various types of valves, actuators, and positioners is shown in Figure 1-1.

Now let's discuss what the control valve should do. Referring to Figure 1-2, which shows a somewhat simplified process control loop diagram, we see three important function blocks above the control valve symbol.

The first one is control valve gain, *Kv*. This is determined by the "installed flow characteristic" of the valve (quite different from the characteristic shown in the vendor's catalog). *Kv* tells you how much the flow through the valve is changing per a given signal change. For input see Chapter 8.

The second block shows the control valve dead time, TD_v. This is the time it takes before a valve moves following a controller signal change. This is usually determined by the valve and actuator friction but may include time lags due to long pneumatic signal transmission lines and the time to build the pressure up in a diaphragm case, for example.

Finally, the third block shows the time constant of the valve, TC_v. This is simply related to the stroking speed of the actuator or actuator/positioner combination (see Chapter 9), i.e., how fast the valve is responding to an upset in your controlled variable. All these function blocks interact, and each one should be considered in evaluating a control valve application.

The "ideal" valve should have a constant gain throughout the flow range, i.e., a linear "installed" flow characteristic, no dead time with packing tightened, and a time constant that is different from that of the process by at least a factor of three.

Need I tell you that there is no such thing as the "ideal" control valve? So let's attempt to develop a workable compromise.

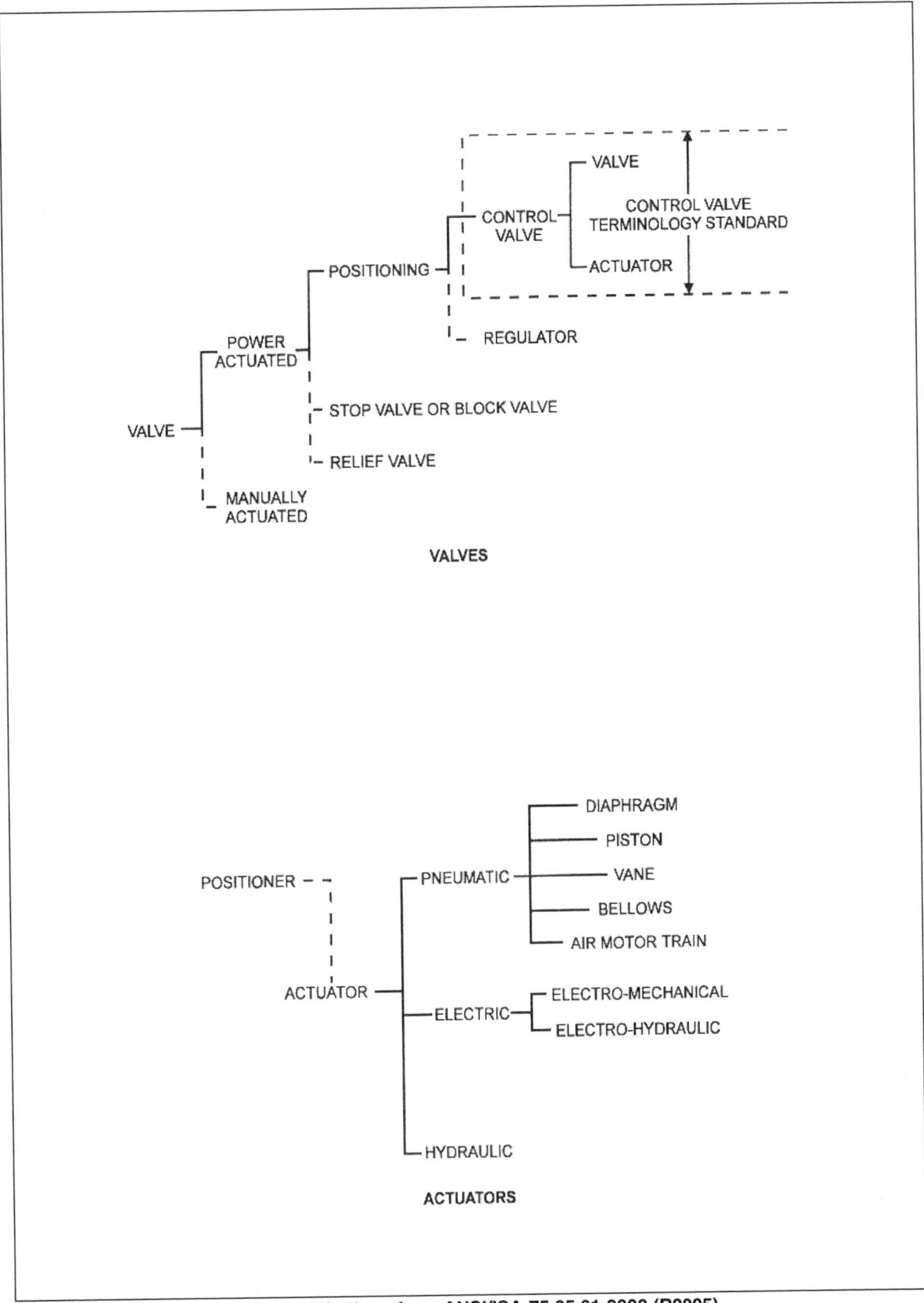

Figure 1-1. Basic control valve terminology from ANSI/ISA-75.05.01-2000 (R2005).

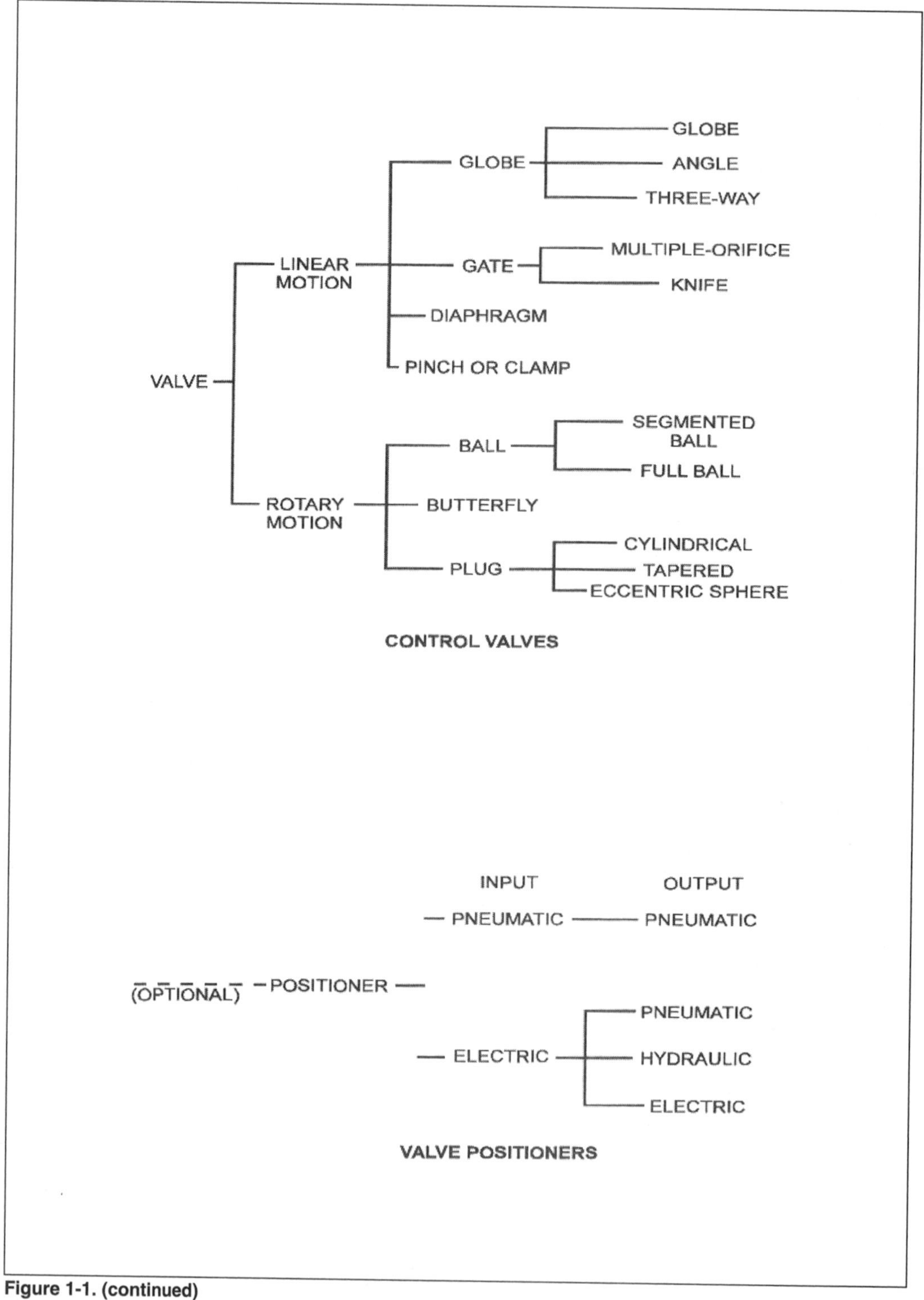

Figure 1-1. (continued)

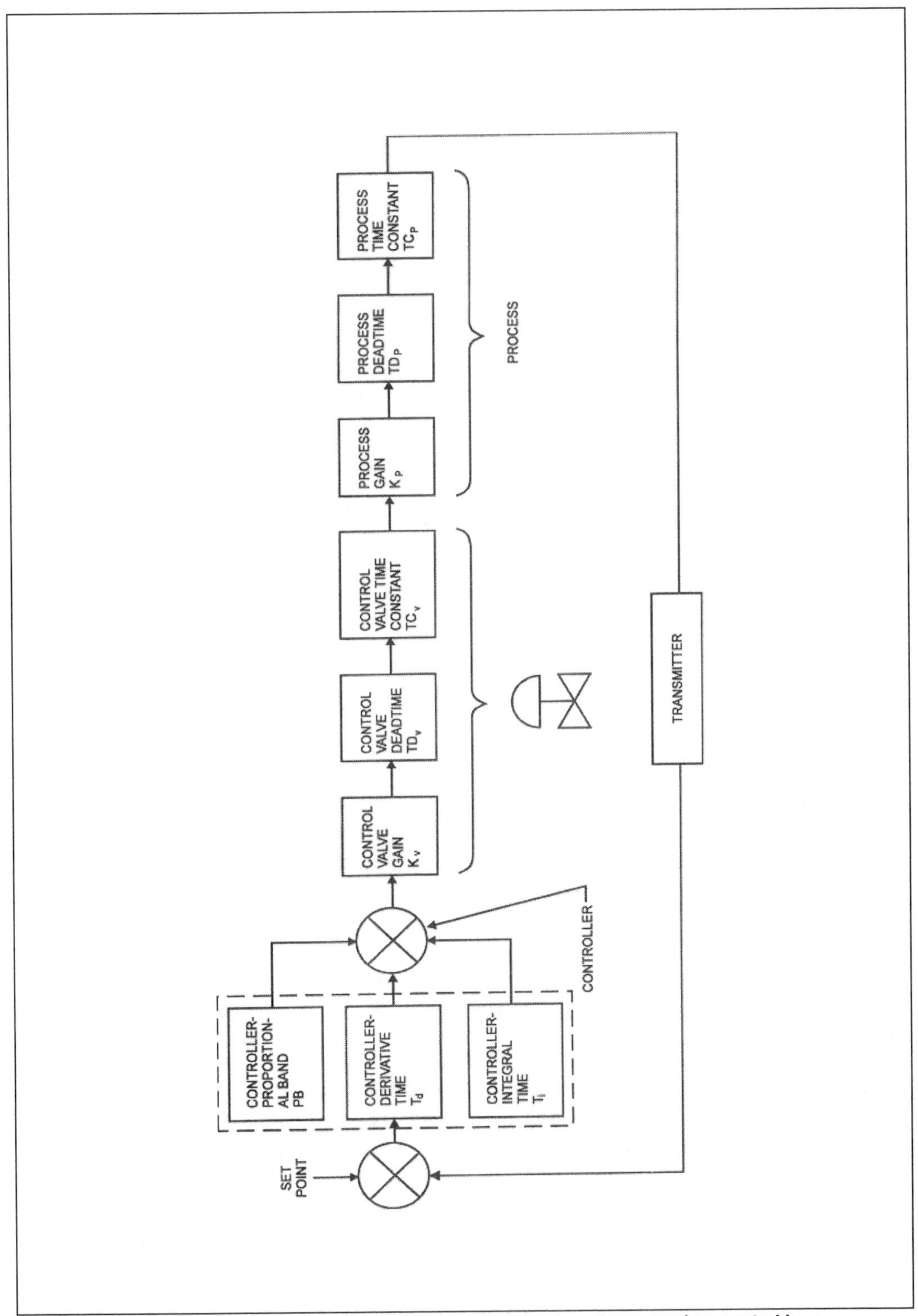

Figure 1-2. Schematic block diagram of controller, control valve, and process in a control loop.

WHAT TO LOOK FOR IN A GOOD CONTROL VALVE DESIGN

Besides the obvious, such as good quality workmanship, correct selection of materials, noise emission, etc., special attention should be paid to two areas:

- Low dead band of the actuator/valve combination (with tight packing).
- Tight shutoff, in cases of single-seated globe valves and some rotary valves (if required).

The prime concern of an operator of a process control loop is to have a loop that is stable. (Nothing makes people more nervous than a lot of red ink and scattered lines on a strip of paper from a recorder.) The final control element will influence the stability of a loop more than all the other control elements combined.

The biggest culprit here is "dead time."[1] This is the time it takes for the controller to vary the output signal sufficiently to make the actuator and the valve move to a new position. What we are talking about here is that the dead time, T_{Dv}, is the time it takes for the pneumatic actuator to change the pressure in order to move to a different travel position. It is most commonly related to the "dead band"[1] of the actuator/valve combination, or, in case a positioner is used, the dead band of the valve divided by the "open loop gain"[2] of the positioner plus the positioner's dead band (dead band keeps the valve from responding instantly when the signal changes, which, in turn, causes dead time). The valve itself should never have a dead band of more than 5% of span, that is, less than 0.6 psi for a 3 to 15 psi signal span or 0.8 mA for a 4-20 mA signal. The positioner/valve combination should have no more than 0.5% of signal span. Ignoring process dynamics, a positioner may, therefore, improve matters by an order of magnitude. However, positioners can raise havoc with the dynamics of a control loop (see Chapter 9). The ideal valve is still the one in which the operating dead band with tight stem packing is less than 1%. There you have maximum stability without the extra cost and complication of having to use a valve positioner. Unfortunately, the majority of valves can not operate without a positioner. Luckily, modern positioners have electronic tuning capabilities that can help to drive the dead band down (see Chapter 9).

The very first control valves recognized the value of low dead band. The valve in Figure 1-3, for example, features ball-bearings around the actuator stem in order to avoid twisting the spring and thereby reducing friction (circa 1932). Low seat leakage can be very beneficial. First, it saves you the extra expense of a shutoff valve in case the system closes down. (Note: For safety reasons, never rely on the control valve for absolute tight shutoff.) Second, it saves energy. Third, it is an unqualified requirement in temperature control applications in a batch process,

---

1 This is often confused with "hysteresis," a dead band produced by moving the actuator up and down through 100% travel. Hysteresis is not very important in a "closed-loop" system.

2 Usually 50 to 200. Consult vendor's published positioner catalog.

especially when you are supplying a heating fluid to a chemical that can undergo an exothermic reaction.

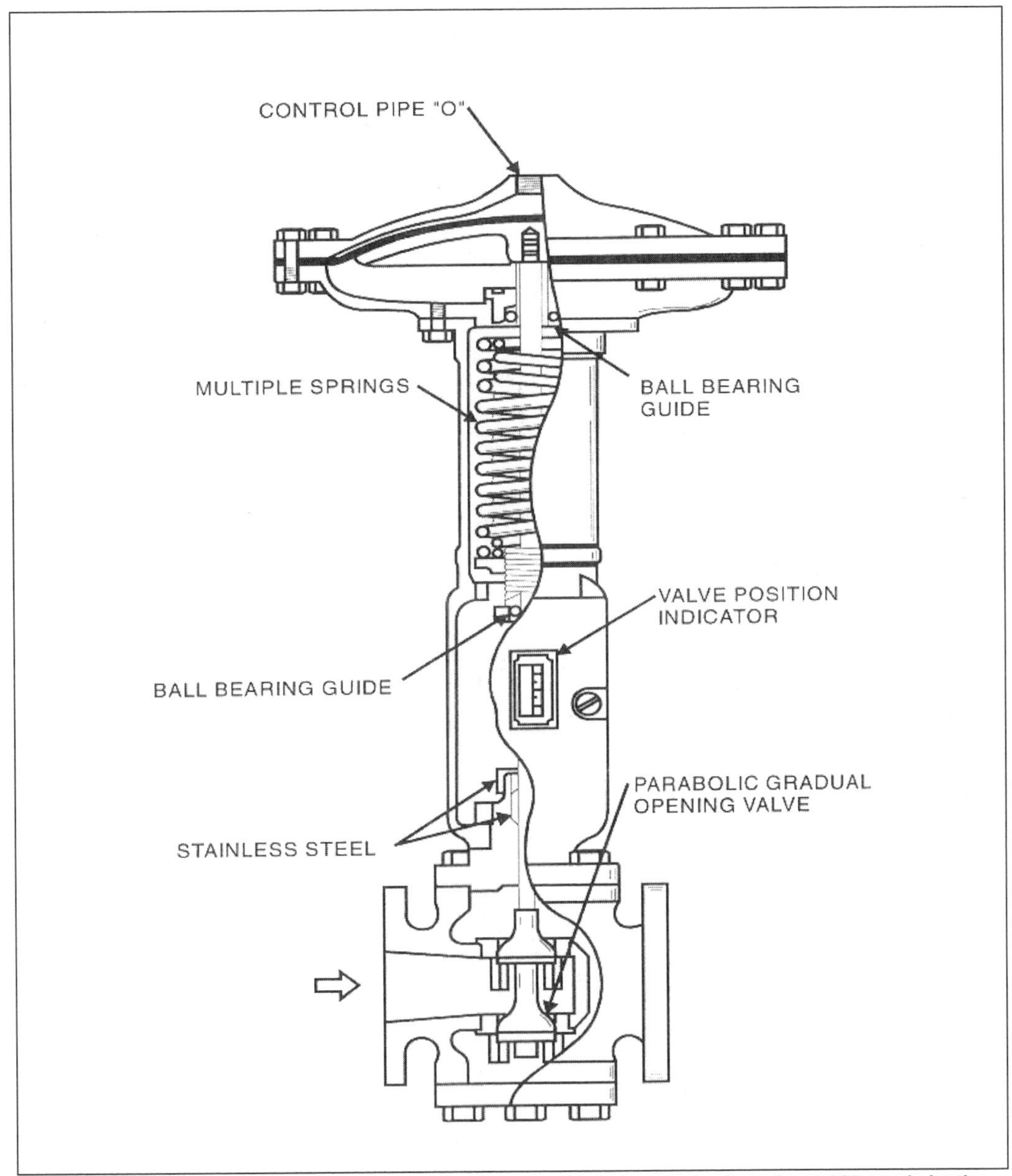

Figure 1-3. Cross section of an early spring-diaphragm actuated control valve, circa 1932, made by the Neilan Co. of Los Angeles, CA. Actuator signal was 2-15 psi.

Another feature is good rangeability. As you probably already know, valves are usually terribly oversized. That is, they operate only at perhaps 30% of their rated C_v under normal flow conditions. Furthermore, it is not wise to operate a conventional valve trim at less than 5% travel.[1] The reason is that the controller might be slightly unstable at the low flow rates, and the actuator, following the sinusoidal output of the controller signal, will push the plug against the seat. When this happens, the actuator and positioner will bleed all the air out, and you end up having a really big dead time, (before the valve gets moving again) upon signal reversal. In short, you have a big mess and no control.

Using the 5% travel limit, we find that the C_v is only about 7% of the rated C_v with a linear characteristic and not much less with most equal percentage characterized plugs. This gives us a useful range for the above-normal C_v of 30% divided by 7%, which is 4.5:1. However, in most loops the pressure drop across the valve always varies inversely with the flow, that is, a low *DP* at maximum flow and a high *DP* with low flow. We, therefore, might find that the actual controllable ratio of maximum to minimum flow is perhaps only 3:1. Incidentally, this ratio is the so-called "installed" rangeability, which, as you can see, is quite different from the "inherent" rangeability shown in a manufacturer's catalog for the same valve. A simple way to remember is that the "inherent" valve rangeability is the ratio of maximum to minimum controllable C_v, while the "installed" rangeability is the ratio of maximum to minimum controllable flow rate in your loop.

REFERENCES

1. McMillan G. K., and Weiner S., *How To Become an Instrument Engineer - The Making of A Prima Donna.* Research Triangle Park, NC: ISA (1987).

---

1 Without a positioner and with high packing friction, such as graphite packing, the minimum travel position should be higher than the dead band caused by the packing friction (usually 7% to 10%)!

2

WHY NOT USE A SPEED-CONTROLLED PUMP?

When the means to electronically control the speed of an electric motor became available in the mid-1970s, a thought immediately occurred: "Why not vary the speed or number of revolutions of pumps in order to control the flow rate of liquids in process control systems?" As a matter of fact, this seemingly ideal solution led to the purchase of a large electronics company by the owners of a major oil company.

Alas, things did not turn out to be that simple. For one thing, modern sizing of control valves, coupled with the introduction of rotary control valves having higher inherent flow capacities, led to the use of pumps with smaller head pressures, hence, less "waste" of energy, i.e., pressure drop across the control valve. Energy seems "wasted" to us because the throttling action of a valve is a thermodynamically nonreversible process and, therefore, increases the "entropy" of the fluid; thus, the energy within the fluid is at a lower, less available state. In practical terms; we consume more electric power than we should.

Unfortunately, as it turned out, reducing the speed of an electric motor driven pump was not as economical as first thought. Following a flow reduction, most of the energy savings at the pump is converted into useless heat within the motor, and more importantly in the variable frequency, and/or variable voltage converters. Most energy saving arguments center around the electric power that can be saved at the motor. However these arguments invariably omit the low efficiency of the current or voltage converters which can be as low as 10% at low motor speed[1,2]. Variable frequency drives have efficiencies of between 70% and 75% at 40% of max. pump speed. Variable voltage drive are even worse, here the efficiency at 40% speed is as low as 20%. This certainly would negate any savings derived from lower pump head! One should only look at the ribbed (for heat loss) converter housings.

Economic arguments for speed-controlled electric motor-driven pumps can be made only in systems where the pump is deliberately oversized, forcing the control valve to reduce perhaps 30% of the available pump head pressure at the rated flow rate. This is simply bad control valve sizing. It should be only 5% to 10% (see

Chapter 5 on Valve Sizing). Keep in mind that in certain pipe lines the control valve pressure drop is as low as 1/2% of the total system pressure! An advantage for speed-controlled pumps exists where the static pressure of the system is relatively low and the flow is relatively constant. However, this is usually a rare occurrence.

Other arguments against electric speed control of pumps[3] are:

- high capital investment (averages twice the price of a valve),
- poor dynamic performance compared to small- and medium-sized valves, and
- higher maintenance requirements.
- limitation on minimum flow due to heat generation in the pumped liquid.

While I might seem to be very negative on this subject, there are indeed applications in which a speed-controlled pump or fan is preferred. For example, if you pump a highly corrosive fluid (and where process dynamics allow). Here the expense of a special alloy valve can be avoided. Other cases can be made for dangerous fluids such as molten metal, etc. Speed control can improve control loop performance due to the absence of valve dead band on applications such as boiler pressure control (fan speed control of forced combustion air).

Sometimes the combination of pump speed adjustment (either electronically or by mechanical means such as adjustable pulleys, for example) and a control valve can be beneficial in order to match the pump head to ideal process requirements. Such systems are used, for example, in laminar flow heat exchange systems in chip production plants.

Quite another argument can be made if the pump speed control is through energy sources other than relatively expensive electricity. A case can therefore be made for the diesel engine-driven pumps used to control pressure in oil pipe lines or for utilizing steam in a power plant to feed turbine-driven feed water pumps replacing the former control valves.

REFERENCES

1. Baumann, H. D., "Control Valve vs. Variable-Speed Pump," *Chemical Engineering,* June 29: 81-84 (1981).
2. Lipták, B. G., Process Control and Optimization, Vol. 4, pp. 1465, Taylor & Francis, Boca Raton, FL (2005).
3. Wood, W. R., "Beware of Pitfalls When Applying Variable- Frequency Drives," *Power,* February: 47-49 (1987).

3

WHAT VALVE TYPE SHALL I CHOOSE?

SELECTING THE RIGHT VALVE TYPE

While performance is certainly number one in our selection criteria, cost cannot be ignored since we all must live within our budgets. After having reviewed a number of bids/proposals, you will soon discover that rotary-type control valves have a definite cost advantage in sizes above 2 inches (50 mm), excluding high-pressure and special-purpose valves. They tend to have less packing leakage if designed properly (see Chapter 14). This then narrows our focus towards globe valves for general service applications in sizes 1/2 in. (12.5 mm) to 2 in. (50 mm) and eccentric rotary plug, or ball valves in sizes above 2 inches (50 mm). Butterfly valves could be considered above eight inches. If you select a globe valve, then try to specify one with a bolted-on bonnet to facilitate maintenance. Bronze or alloy valves may have threaded bonnets.

In order not to box yourself in price-wise, try not to specify what valve type should be quoted, but rather specify what features *do not work* in your given application. For example, if you have to control crude oil, specify that "no cage-guided trim" is allowed (the solid particles may jam the plug guide!). If you handle hazardous fluids, specify "no threaded end connections," and so on. Specify tight shutoff only if really necessary; it may cost extra money (see Chapter 14).

Only you know the corrosive nature of your process fluid best; therefore, specify stainless steel, alloy, or plastic as body or trim material (consult corrosion table). However, for installations in process plants, you may want to refrain from using all-plastic valve bodies. Remember, anything that can be used by a maintenance person as a step ladder will be used as such!

When using a plastic-lined metal valve body, make sure the surrounding housing is pressure-tight and that a secondary stem packing and telltale connection are provided to protect you against liner or plastic bellows failure. Bellows made of tetrafluoroethylene (TFE) — Teflon® plastic — have particularly limited life cycles and are permeable to chlorine.

Valve end connections may be threaded in sizes 2 in. and below for non hazardous fluids but should be flanged for the handling of chemicals. High-pressure steam valves usually are welded into the pipe, as are some aseptic valves in bioprocessing systems.

While it is usually "OK" to have your favorite valve vendor select the valve type (which saves me from adding all the usual catalog pages to this volume), base your final judgment on the criteria mentioned herein. The maintenance and past experience of a given type of valve is always an important consideration.

One of the most embarrassing things that can happen is seeing a valve you chose six months earlier on the trash heap and then listening to unflattering remarks made by the maintenance foreman.

By far, the most often encountered causes of premature valve failures are:

- cavitation damage, (usually high pressure drop with water).
- wire drawing of trim, (cavitation or flashing of water, condensate).
- erosion of trim, (embedded solids in fluid).
- plug vibration, (unstable flow pattern with liquids).
- corrosion, and (corrosive fluids).
- excessive stem packing leakage (valve/actuator may be cycling).

The first can be avoided by selecting a low-pressure recovery trim (high F_L) by installing a secondary flow restrictor or by choosing a better valve location. The second is usually encountered with wet steam (condensate actually). Here the carbon steel of a valve housing can get eaten away underneath the stainless steel seat ring if there is no gasket or weld seal. The moral: Always specify chrome-molybdenum alloy or stainless steel as body material where saturated water is encountered or where you suspect steam may condense (practically any process steam line that gets shut down at night or on weekends). For the third problem, specify hardened trim or use angle valves. There are, in my book, only two cases when you should use an angle valve: (1) for flashing condensate at the inlet of a condenser and (2) for erosive fluids. In both cases, the flow should be in the direction to close the plug (check actuator for stability). For erosive service, use a reduced trim and a long, straight, (ideally) oversized downstream pipe to avoid particle impingement on the pipe interior. The fourth problem can sometimes be cured by reversing the flow direction; otherwise, increase the stem diameter or reduce the weight (mass) of the valve plug. The fifth problem can be dealt with when selecting materials with the aid of a good handbook (see Chapter 13). For hints in solving the problem of excessive stem packing leakage, see Chapter 14.

Larger **Globe Valves** may have a pressure balanced plug and could be cage guided. Figure 3-1 is a typical example (of an improved) cage valve design. Cage valves have a number of advantages; they are easier to maintain since removal of a seat ring is relatively easy compared to the removal of a threaded seat ring in a single seated globe valve, especially after years of service. Pressure balanced plugs greatly reduce the stem forces and can save considerable amounts of money by the ability to use smaller actuators. However, there can be some problem areas one should consider: first, cage-guided plugs are susceptible to entrained particle leading to friction and stickiness; secondly, there can be a problem with thermal expansion of the cage on high temperature service that could result in crushing (leakage) of the cage gaskets; thirdly, the cost of a cage valve body (and its weight) is higher than that of a single-seated globe valve due to the heavy bonnet bolting. Finally, most cage plugs (pistons) are sealed against the gage bore using piston rings (either metal or plastics), this can cause considerable friction during operation, leading to dead bands. The positioner may have to be de-tuned in order to prevent cycling of the actuator (valve instability). Such valve – actuator instability should not be confused with instability (hunting) of the control loop, although both may inter-act.

The idealized cage valve in Figure 3-1 shows ways how these inherent disadvantages of cage valves can be overcome. Most intriguing is the concept of dual stage throttling. Here, the plug instead having a conventional contoured surface (which determines its flow characteristic) the contoured surface is now part of a radially expanding inner cage wall, located just above the valve seat. The clearance between the tip of the plug (lower letter "A") and the contoured inner cage wall now becomes the first throttling stage. At the same time, we see a second protrusion of the plug (upper letter "A") starts to expose a portion of a window cast into the cage. The resultant opening thus becomes the second stage of this dual throttling element. Fluid will pass past the valve seat and lower tip of the plug into a concave indentation and from there, restricted by upper point "A", into the open window areas, and finally exiting the valve body."

Smaller globe valves usually feature single-seated designs as shown in Figure 3-2. These are much simpler and can use smaller actuators due to low stem forces due to smaller orifices and generally lower pressures. This also allows for smaller stem diameters hence lower packing friction (less dead band). Many of these valves can operate without a positioner if operated using a 3-15 psi pneumatic signal.

Rotary control valves fall into two broad categories:

- Those designed exclusively for modulating control;
- On-off valves modified by the addition of an actuator and positioner (also called "automated valves").

In the first category are many features commonly seen in globe valves such as adjustable stem packing, the option of a hard-faced trim, extension bonnets, and, most important, a backlash-free connection between the rotating "closure member" and the actuator stem. Some more refined types even have two-stage pressure reduction at low travel and field attachable low noise and anti cavitation inserts (see Figure 3-3).

The second category, while admittedly less expensive, can be perilous. Here the main concern should be the backlash or play between:

- the ball (in case of a ball valve) and the valve stem tang,*

* The flattened portion at the end of a valve stem.

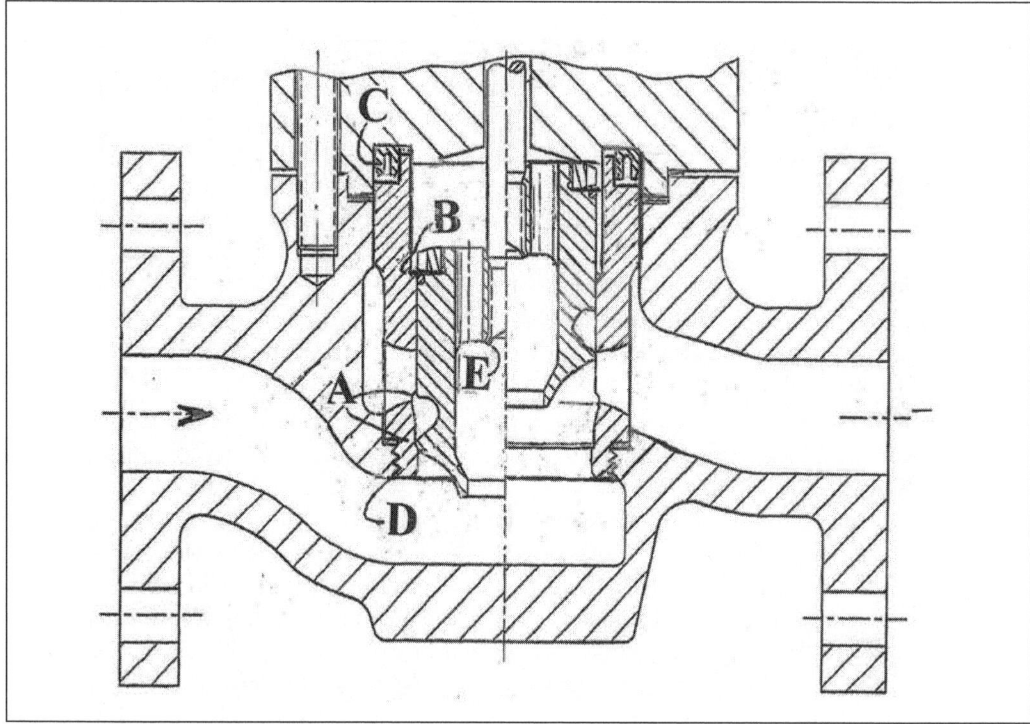

Figure 3-1. Features a modern cage type control valve body sub-assembly. The valve plug allows for double stage throttling (A) for increased cavitation resistance and high rangeability. Friction causing sliding seals are avoided by the use of a flexible top seal (B) allowing for balanced forces and for tolerating machining tolerances and thermal expansion. A similar feature is used to seal the bonnet against the cage by using a compressible top seal (C). The cage is screwed into the housing (D) eliminating the heavy bolting required to compress two gaskets on conventional cage valves. Finally, the stem connection to the plug is staked (E) instead of using a pin which may become loose. [Reference: US Patents 6,701,958 ; 6,766,826 ; 7,083,160 (FISHER CONTROLS INTERNATIONAL)]

- the valve stem and the actuator stem tang, and
- the positioner feedback connection and the actuator stem tang.

Here is a little example to illustrate the above mentioned problems: Assuming the valve has a 1 in. stem diameter and 0.015 in. of tang clearance, a little mathematics will prove that the actuator will have to move 2.5° or 2.8% of 90° travel before the ball will move in the opposite direction. This is acceptable in some applications, but not for critical control.

Another area of concern is "break away" friction with some ball or butterfly valves. This is overcoming the static friction of the ball or disk in order to open the valve; a problem especially irksome on so-called "high-performance" butterfly valves. Upon start-up, it may happen that the actuator is so charged up with air pressure that the valve may fly almost wide open (after breaking out of the seat) before the positioner can catch its breath and move the valve back to the desired position. A good rule to remember is to limit the normal valve travel of the ball and high-performance butterfly valve to 15% of max. opening, i.e., make sure

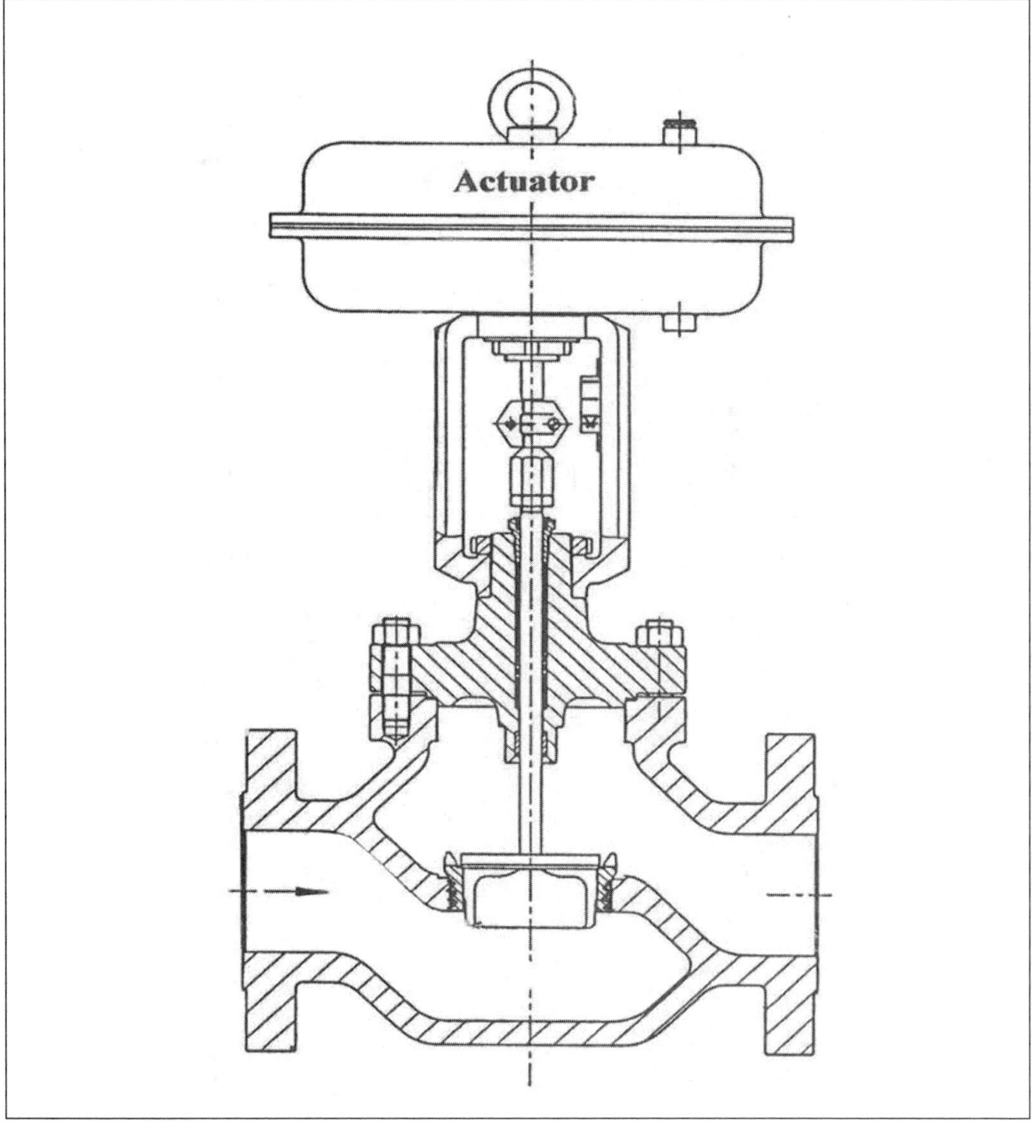

Figure: 3-2. A single-seated control valve featuring a screwed in seat ring and a V-port plug. Plug and seatring are typically made from type 316 stainless steel. Courtesy: SAMSON A.G.

your flow rate does not require a C_v *(flow coefficient)*, which is less than that given for a 15% opening.

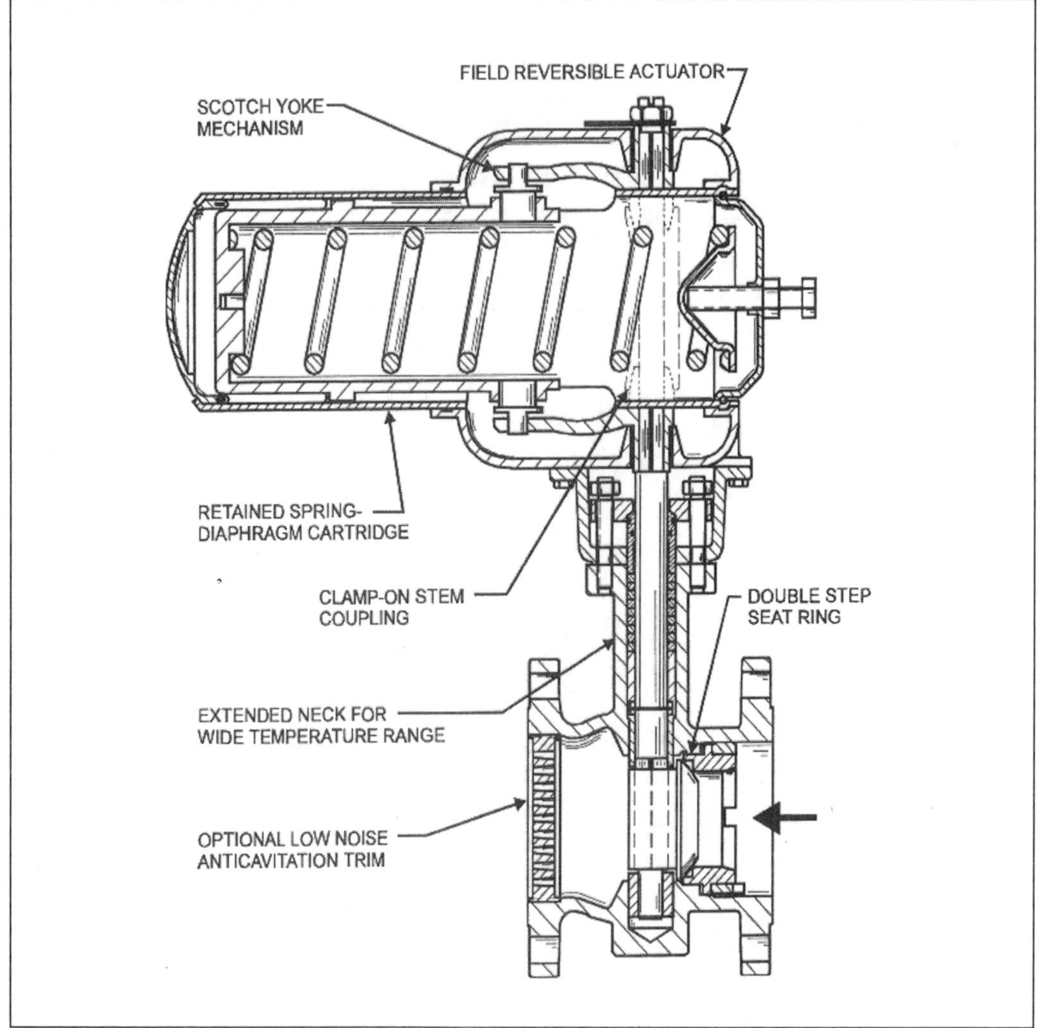

Figure 3-3. Example of rotary valve designed especially for control purposes. Note field-replaceable low-noise insert. (Courtesy of H. D. Baumann Inc.)

CRITERIA FOR VALVE SELECTION: A SUMMARY

Ability to Control Well (Performance)

Is the valve sized properly?

Does the valve have low dead band, sometimes confused with hysteresis?

Will it be able to operate without a positioner (also a cost factor)?

Does it have adequate rangeability?

Is the flow characteristic acceptable (installed gain)?

How is its dynamic performance (frequency response, time constant)?

Installation and Effects on the Environment

Is the face-to-face such that it can be replaced with another make?

How heavy is it? Are pipe supports needed? Is it too tall?

Does it shut off tightly (if required)?

Is it going to be noisy under the given flow conditions?

Would packing leakage produce a hazard?

Is there adequate air pressure (in case of cylinder actuators)?

Can it withstand the corrosive effects of fluid *and* environment?

Would the valve be cavitating under the given flow conditions?

If outdoors, can the valve and positioner stand extreme weather conditions?

Maintainability and Long-Term Cost Control

Is it easy to repair?

What are the costs of spare parts?

Can the positioner handle dirty air (if applicable)?

Can the positioner tolerate a vibrating environment?

How is the actuator protected from corrosion?

Is the stem packing field-replaceable?

Does the valve need to be drained?

Is the positioner able to diagnose the valve?

Only after reviewing these criteria and deciding on the proper valve type should the subject of cost be considered.

4

THE SELF-ACTING REGULATOR, WHY NOT?

The self-acting regulator is a relatively inexpensive, complete control system that combines the process sensor, the controller, and the control valve all in one. Sadly, the market share of this device is dying out from sheer neglect. Properly utilized, regulators could replace control valves in at least 25% of all control loops. However, they are seldom specified in a process loop because of the instrument engineer's unfamiliarity with them. Yet almost every valve positioner comes with its own air set, which is nothing but a self-contained air pressure regulator.

The advantages of regulators,[1] besides cost, are: an excellent frequency response (in virtually milliseconds), very good rangeability, space saving, usually no stem leakage, and they do not require air or electricity to operate. Here, the process fluid itself perhaps the differential pressure across the valve or a temperature differential, provides the necessary energy to operate the valve.

The disadvantages are: fixed proportional band (typically 25% for direct and 5% for pilot-operated models), no reset action, limited sizes and pressure ratings, and limited choice of materials and end connections. The maximum C_v rating is usually about one half that of a control valve, which is still adequate for line-size installations.

An important safety consideration is the fact that spring-loaded, pressure-reducing types of regulators will open fully if the operating diaphragm fails! It is, therefore; important to make sure that the downstream piping can withstand upstream pressure conditions; otherwise, a downstream relief valve that can handle the maximum flow of the regulator must be installed. Spring-loaded, backpressure regulators will close upon diaphragm failure, and a similar backup for protection of the upstream piping system is advised. Nevertheless, when the application calls for simple control of a blanketing gas in a tank, the steam tracing of a crude oil pipe, or other secondary loop applications, regulators should be given serious consideration.[2] For example, pressure reducing regulators are widely used in natural gas distribution systems. They may also be the answer to applications where very fast reaction to a process upset is a must. These are typically so-called back pressure regulators. Here a regulator is acting in a way like a safety relief valve. One typical application is for surge pressure control in pipe lines.

Problems encountered when employing self-contained regulators are: instability under certain flow conditions. This can lead to resonant vibration of the moving parts which, in turn, can create fatigue failure of metal diaphragms.

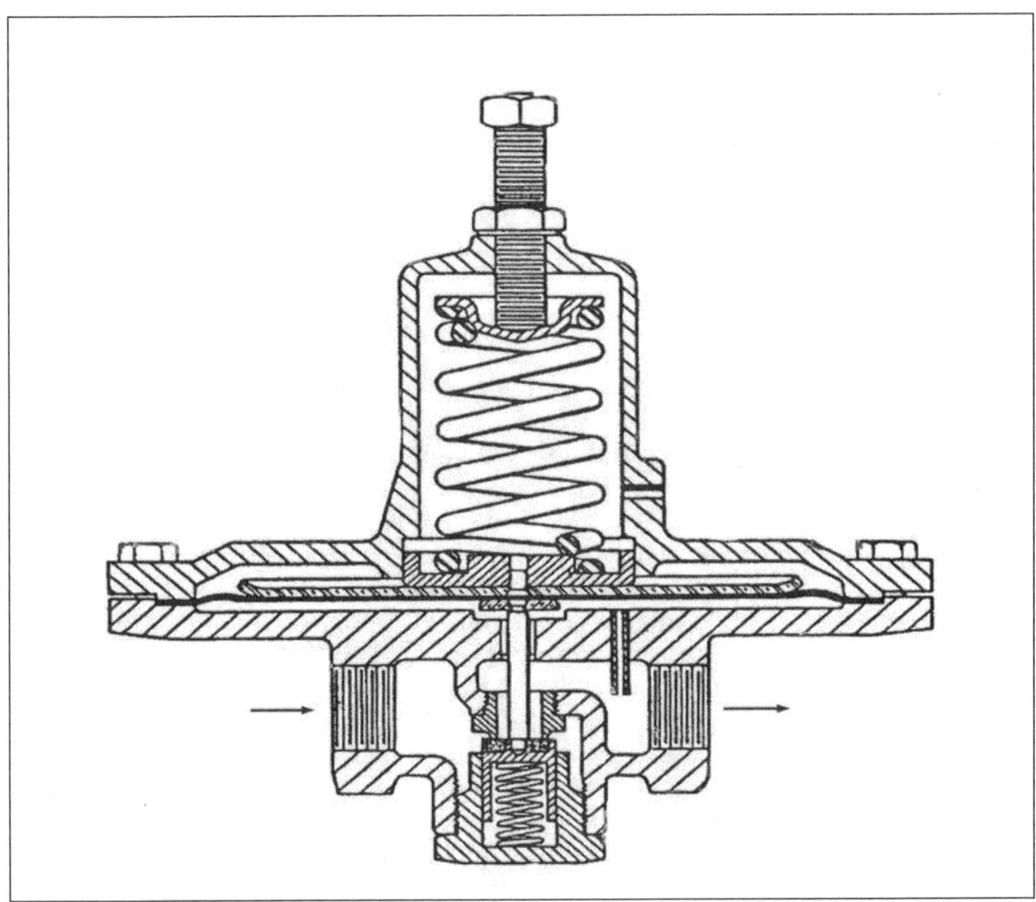

Figure 4-1. A spring-diaphragm operated pressure reducing regulator. Adjustable spring load acting on diaphragm determines desired valve outlet pressure.

REFERENCES

1. Moore, R. L., "The Use and Misuse of Pressure Regulators," *Instrumentation Technology,* March: 52-56 (1969).
2. Baumann, H. D., "Regulator or Control Valve," *ISA Journal,* December: 51-52 (1965).

5

VALVE SIZING MADE EASY

VALVE SIZING

This activity can be defined as picking the right size valve for the job or, in certain instances where a valve size is predestined, the correct trim size.

Do not read on if you have a computer fully programmed with ANSI/ISA-75.01.01 equations.[1] But then again, you may want to know what you are doing.

To help matters, about 60 years ago some clever soul invented a valve sizing coefficient called C_v that combined the flow area of the valve orifice, the contraction coefficient, and the head loss coefficient all in one; thus,

$$C_v = q/\sqrt{\Delta P/G_f} \tag{5-1}$$

$$q = C_v\sqrt{\Delta P/G_f} \tag{5-2}$$

where q is the flow rate in U.S. gallons per minute, G_f is the specific gravity of a liquid at a given temperature (water at 60°F has a G_f of one), and ΔP is inlet minus outlet pressure in psi.

Hence, one C_v equals the flow of one U.S. gallon per minute of water at 60°F and under a pressure drop across the valve of 1 psi. Now the same valve will flow 10 gpm at a pressure drop of 100 psi since the flow is proportional to the square root of ΔP, thus following the laws of fluid mechanics. This equation is worthwhile memorizing since you will use it on at least 80% of all sizing problems for liquids. Unfortunately, not all in life is so simple; therefore, we have to recognize two limitations in the use of the above equations.

First, they only apply to turbulent flow, i.e., Reynolds numbers above about 5000. However, you can ignore this limit on required C_v numbers above 0.1 and when the kinematic viscosity is below 40 centistoke (90% of all liquids are). There will be more on this subject—required reading only if you control tar or equally sticky substances and polymers.

Second, in some conditions funny things occur inside the valve which seems to block the flow passages. What really happens is that at a certain ratio of pressure drop to inlet pressure defined by a pressure recovery factor $F_L{}^2$ the flow becomes choked at the orifice. (F_L originally was called *the* "critical flow factor,"[2] C_f and was used as such.) In the case of liquids, this is caused by the evaporation of some of the liquid (also see Cavitation in Chapter 14), and in the case of gases, the onset of sonic velocity. We could not care less about all this were it not for the fact that now the flow is independent of the downstream pressure P_2. Now the flowing quantity can only increase if you use a higher inlet pressure.

To avoid embarrassing errors, you should first start to determine if you have critical (choked) flow or not using the equations below (this also helps to discover any cavitation or noise problems you may have to worry about).

Note, use $(F_{LP} / F_P)^2$ instead of $F_L{}^2$ in the equations below whenever you have a valve between pipe reducers where

$$F_{LP} = 1/\sqrt{\frac{1}{F_L^2} + \left(\frac{C_v}{30d^2}\right)^2\left(1 - \frac{d^4}{D^4}\right)} \tag{5-3}$$

here d = valve size (in.) and D = pipe size (in.).

For F_P, see subchapter on Pipe Reducers.

For Liquid Service

SUBCRITICAL FLOW	CRITICAL FLOW	ΔP_s = Maximum ΔP for sizing.
Is if ΔP is less than $F_L{}^2\ (\Delta P_s)$	Occurs when ΔP is more than $F_L{}^2\ (\Delta P_s)$	Use: $P_1 - P_v$ when outlet pressure is higher than vapor pressure.

Volumetric Flow

When outlet pressure is lower or equal to vapor pressure use: $\Delta P_s = P_1 - [0.96 - 0.28\ (P_v / P_c)]$

$$C_v = q\ (G_f / \Delta P)^{0.5} \qquad C_v = q / F_L\ (G_f / \Delta P_s)^{0.5} \tag{5-4}$$

Flow by Weight

$$C_v = \frac{W}{500\sqrt{G_f \Delta P}} \qquad C_v = \frac{W}{500 F_L \sqrt{G_f \Delta P_s}} \tag{5-5}$$

G_f = specific gravity @ flowing temperature (water = 1 @ 60°F)

P_1 = absolute inlet pressure, psia

P_2 = absolute outlet pressure, psia

ΔP = $P_1 - P_2$

P_c = pressure at thermodynamic critical point, psia—water is 3206 psia

P_v = vapor pressure of liquid at flowing temperature, psia — water = 0.4 psia @ 70°F

q = liquid flow rate, U.S. gpm

W = flow in pounds per hour

For Gas and Steam Service

Surprising as it may sound, the basic C_v* equation can also be applied to gases and steam. The difference here is that the density of the gas changes with ΔP, and, since this is a gradual process, the relationship $\sqrt{\Delta P}$ to flow is no longer linear but curved. To its credit, the ISA Sizing Committee for ANSI/ISA-75.01.01 has labored hard to predict the exact shape of this curve with the aid of an expansion factor[3], Y, where

$$Y = 1 - (x/3F_k X_T) \qquad (5\text{-}6)$$

Note that Y cannot exceed 1 and that it cannot be less than 0.67 since the latter is exactly the density of the gas, i.e., 67% of the inlet density when the valve is choked (when it sees sonic velocity in the valve orifice).

In the above equation,

x = $\Delta P/P_1$ (Note: P_1 *or* P_2 is *always* absolute pressure, i.e., gage psi plus 14.7)

F_k = the ratio of the specific heat k of a gas (air = 1.4), $Fk = k$ of gas/1.4

X_T = pressure drop ratio. This is the $\Delta P/P_1$ at which there is no longer any flow increase.

Note, that X_T includes some flow increase *after* the onset of choked flow due to changes in the size of the flowing jet (vena contracta).

Now the required flow coefficient becomes:

$$C_v = q(G_g T_1 Z/x)^{0.5}/(1360 P_1 Y) \qquad (5\text{-}7)$$

* There is evidence that the C_v number on certain valve types, such as angle valves, certain rotary valves, and double-ported globe valves is different between liquid and gases, but we shall ignore this.

Or

$$q = 1360 C_v P_1 Y / \sqrt{G_g T_1 Z / x} \qquad (5\text{-}8)$$

where:

- q = flow in scfh
- Gg = gas specific gravity (air = 1), also ratio of molecular weight of gas to that of air, (air = 28.9)
- T_1 = absolute upstream temperature °R (°F + 460)
- Z = compressibility factor (ignore if your pressure never exceeds 2000 psia; otherwise, consult handbook tables.)

Now, if this begins to tax the capabilities of your hand-held calculator, you may ignore all of the above and use the simplified version below.[4] (The penalty is a maximum sizing error of ≈8% at the transition from nonchoked to choked flow and not worth the trouble.) It is also strongly recommended that you use these simplified equations to make sure your computer-generated answers are at least in the ball park!

SUBCRITICAL FLOW When ΔP is less than $F_L^2 (P_1/2)$	CRITICAL FLOW When ΔP is more than $F_L^2 (P_1/2)$

Volumetric Flow

$$C_v = \frac{Q}{963}\sqrt{\frac{G_g T}{\Delta P(P_1 + P_2)}} \qquad C_v = \frac{q\sqrt{G_g T}}{834 F_L P_1}$$

Flow by Weight

$$C_v = \frac{W}{3.22\sqrt{\Delta P(P_1 + P_2) G_g}} \qquad C_v = \frac{W}{2.8 F_L P_1 \sqrt{G_g}}$$

For Saturated Steam

$$C_v = \frac{W}{2.1\sqrt{\Delta P(P_1 + P_2)}} \qquad C_v = \frac{W}{1.83 F_L P_1}$$

For Superheated Steam

$$C_v = \frac{W(1 + 0.0007 T_{sh})}{2.1\sqrt{\Delta P(P_1 + P_2)}} \qquad C_v = \frac{W(1 + 0.0007 T_{sh})}{1.83 F_L P_1}$$

where:

C_v	=	valve coefficient
F_L	=	pressure recovery factor
G_g	=	gas specific gravity (air = 1.0)
P_1	=	absolute upstream pressure, psia
P_2	=	absolute downstream pressure, psia
ΔP	=	pressure drop $P_1 - P_2$, psi
q	=	gas flow rate at 14.7 psia and 60°F, scfh
T	=	flow temperature, °R(460 + °F)
T_{sh}	=	steam superheat, °F
W	=	flow rate, pounds per hr

NOTE: When a valve is installed between reducers, use $C_v F_p$ instead of C_v in capacity tables (see subchapter on Pipe Reducers).

OTHER METHODS

$$C_v = \frac{C_g}{40F_L}; \quad C_v = \frac{C_s}{2F_L}; \quad X_T = 0.84F_L^2$$

$$F_L = \sqrt{K_m}; \quad F_L = C_1/40$$

NOTE: X_T is not strictly related to F_L. While X_T gives you the maximum gas flow, which may happen at orifice pressure ratios P_1 / P_o of 3:1 or more, i.e., above sonic velocity, F_L^2 tells you the onset of sonic velocity, i.e., where $P_1 / P_{orifice} \approx 2$, and $P_1 / P_{2\ sonic} = P_1 / (P_1 - 0.5\ F_L^2)$.

WHAT ARE THE RIGHT FLOW CONDITIONS?

Using a computer and sizing the C_v number to the fifth place after the decimal point is all well and good, but only if you have the right process conditions. First of all, the valve should be sized to control the maximum flow rate your process is designed for. But what is this exactly? How about emergency conditions? On the other hand, should the valve fail open, would the resultant flow be more than the downstream pressure relief valve could handle? Then again, what is the correct inlet pressure? Can you use the head pressure of the pump from the manufacturer's published curve? (If you do, use the head pressure corresponding to the maximum flow the valve has to pass, and don't ignore the static head at the pump's location.) What is the head loss or pressure drop across a heat exchanger next to the valve? How about line losses? All these are questions that beg only vague responses from your process design people. They don't know either and have to rely on the input from other vendors. The result is usually guesswork with ample safety factors thrown in. Therefore, don't be too concerned about the accuracy of sizing equations. Ninety-nine out of one hundred times, the valve is too large for the job anyhow (see Chapter 8 on equal percentage trim). The remaining one percent that is undersized comes about because someone made a decimal point error in the C_v calculation.

If you want to be clever and save your company energy and money in the bargain, have the process people *ask you* for the valve pressure drop when they design the system (see Chapter 18). Now you are in the driver's seat. Here are the very scientifically derived rules of thumb you should memorize: For ΔP valve sizing, use the higher of 5% of the total system pressure (i.e., static head plus pump head at maximum flow, boiler pressure, etc.), or 5 psi for rotary control valves, or 10 psi for globe valves.

You may wonder why I picked 5 psi for rotary control valves. Well, it just so happens that the flow rate (either liquid or gas) generated by a 5 psi pressure drop through a wide-open control valve with a C_v divided by the valve diameter squared of 17* is all that a pipe (the same size as the valve) can handle, allowing for the customary maximum line velocities of 15 ft/*s* for liquids or 150 ft/*s* for gases [5] as you can see from Figure 5-1. Higher ΔP's mandate a valve smaller than the pipe size and where reducers are required, or when reduced trim is selected.† In any case, you are throwing away dollars for pump, compressor, or boiler horse powers if you assign too much pressure drop to the valve (see chapter 18).

Of course, the above trick can be used in reverse. You may be requested to replace an old steam reducing valve, vintage 1948. All the records are lost, and the only thing you know is the boiler pressure and the valve and pipe size. Say the boiler pressure is 95 psig (110 psia) and the pipe size is 1-1/2 inches. First, look up the maximum flow rate of a 1-1/2 in. steam pipe at, say, 100 psig pressure. You will find it to be approximately 1800 lbs/hr (see Table A-2 in Appendix A). Next, you assume the flow to be choked, i.e., P_2 is less than one-half P_1. In any case, people didn't care much about energy savings in 1948! Now calculate the C_v required, which is about 10.5. So go ahead and select a 1-1/2 in. globe valve with a rated C_v of about 22. The worst that can happen is that the outlet pressure is 80 psig instead of below 50 psig. No problem. Now the C_v required is 16, which is still OK. Therefore, when in doubt, use a full-trim size if flow conditions are unknown, except in special applications such as pH control.

PIPE REDUCERS

One of the competitive strategies of some valve manufacturers is to offer a control valve that has a much higher C_v rating than somebody else's valve. These are usu-

* A butterfly valve at 60° opening, for example.

† For example, if a 4 in. butterfly valve is installed in a 6 in. pipe, then at the same $17 \times d^2 = 272$ C_v, the valve pressure drop increases from 5 psi to 25 psi with 15 ft/s fluid velocity in the 6 in. pipe.

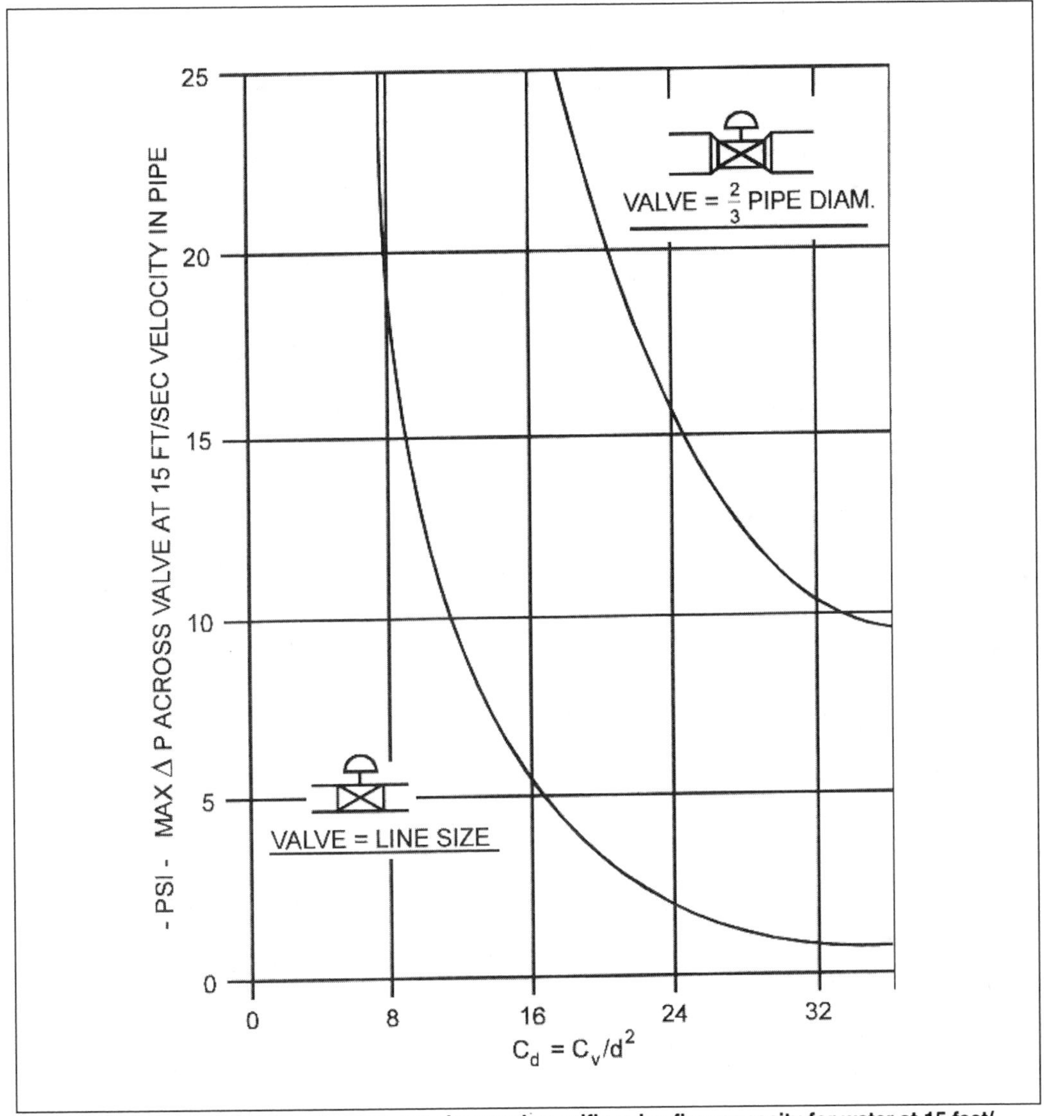

Figure 5-1. Relationship between pressure drop and specific valve flow capacity for water at 15 feet/ second velocity in pipe, *d*= valve diameter (inches). This works also well for gas or steam at 150 ft/sec pipe velocity.

ally oversized globe valves with reduced flange connections, for example, a 1-1/2 in. flanged globe valve with a 2 in. orifice size. While there is nothing wrong with this concept (after all, you are getting more weight for your dollars), it nevertheless forces you into using pipe reducers, since with a large C_v your valve can be smaller than the attached pipe. So one thing you have to do is evaluate the extra cost and space requirement. You may also need reducers when you buy rotary control valves, which, due to having less torturous bodies, have inherently more C_v — although with higher ΔP's, this cancels out because of low F_L numbers (choked flow is dependent on the product of C_v times F_L!).

A word of caution: Never choose a valve that is less than half the pipe size. Too much stress may be created by the movement of the pipes. The only two times in my life when I was genuinely scared were (1) witnessing a live test with steam on a newly designed steam dump valve for nuclear power and (2) seeing a 1 in. steam makeup valve welded into a 10 in. schedule 80 pipe, reducing 1800 psi steam to 10 psig!

There is a good reason why I put this discussion at this particular point in the chapter. Namely, reducers do not affect you that much. Simply remember that the combined flow coefficient of the *valve and reducer* is $C_v\,F_p$ where F_p is the piping geometry factor (which I used to call R,[6] a term which was used by one manufacturer). Some vendors have published tables showing the reduced C_v =($C_v\,F_p$) due to the pressure losses caused by the pipe reducers. *Make sure the C_v or $C_v\,F_p$ value of the valve you select is at least 10% higher than the C_v number calculated with the above equations.* For globe valves at rated C_v's of 10 d^2 or less (d is valve size in inches), F_p is practically one, so it can be ignored. Things are more critical with butterfly or ball valves where F_p can be as low as 0.6!

Here is a quick guide to F_p factors. First, select a valve for say ½ the diameter of the pipe, look up the max. C_v of this valve from your vendor's catalog. Now divide the valve catalog C_v by the valve diameter d (inches) squared, then read F_p:

Table 5-1. F_p Values for Valves, $D/d = 2$

C_v/d^2	10	15	20	25	30	35	40	45
F_p	0.96	0.91	0.85	0.79	0.74	0.68	0.63	0.59

D = pipe diameter (inches)

Next you calculate the required C_v from the given process date and, finally divide this calculated C_v by F_p. This is the final C_v that should be stated on your purchase order. *Note: You may have to start over if the originally selected valve was not large enough, although this is rare.*

For example. You have a 4" pipe. Next select a 2" eccentric rotary plug valve with a catalog C_v of 60. C_v/d^2 is 15. This gives an F_p factor of 0.91 from the above table. Now you determine that the process data call for a required C_v of 45. Dividing 45 by 0.91 requires a min. rated C_v of 49.5. This is no problem, since the catalog C_v is 60.

CORRECTING FOR VISCOSITY

Since this is a procedure not often encountered, I put it last. Ignore this unless (a) you have a liquid with a viscosity exceeding 40 centistokes (or = 40 centipoises for liquids), or (b) you need really small valves with a C_v of less than 0.1.

Similar to the reducer correction, we have to increase the C_v from that calculated with the standard sizing equations in order to make up for the additional friction caused by the stickiness of the stuff passing through the valve. This is expressed by the "valve Reynolds number factor," F_R. The valve C_v you *really* need = C_v originally calculated/F_R. More precisely,

$$C_{v\,corrected} = \frac{C_{v\,turbulent}}{F_R} \tag{5-9}$$

where:

$C_{v\,turbulent} = q\sqrt{G_f/\Delta P}$ (see valve sizing formulas)

$$Re_v = \frac{17300 F_d q}{C_{vR} F_L} \quad (Re_v = \text{valve Reynolds number}^7) \tag{5-10}$$

q = flow rate, gpm

C_{vR} = rated C_v of valve selected (may have to be increased if $C_{v\,corrected}$ is larger.)

F_L = pressure recovery factor

v = kinematic viscosity, centistokes ($10^{-6} m/s^2$) = centipoises/G_f. For typical v values, see Tables 5-2 and 5-3.

$$F_d = \text{valve style modifier,} \quad \frac{d_H}{d_o\sqrt{N_o}};$$

where:

d_H = hydraulic diameter of a single flow passage (4 x area/wetted circumference), inch

d_o = diameter of equivalent circular orifice of a single flow passage

$$d_o = \sqrt{4C_v F_L/38 N_o \pi}\text{, inch}$$

N_o = number of equal flow passages

Typical Values for F_d: (from reference 8, also refer to Table 14-2)

For Globe valves ($C_v > 0.1$):

Single-seated, parabolic plug = 0.46

Double-seated, parabolic plug = 0.32

Butterfly valve = 0.57

Segmented ball valve = 0.98 (wide open only) and 6 in. and above

Eccentric rotary plug valve = 0.42

NOTE: The above F_d values agree with those listed in ANSI/ISA-75.01.01 (IEC 60534-2-1 Mod)-2007.

Example: Bunker C Oil @ 120°F; $P_1 = 100$ psig; $P_2 = 75$ psig; $G_f = 0.97$; $q = 26$ gpm; v = 750 centistokes;

$$C_{v\ turbulent} = 26\sqrt{0.97/25} = 5.12$$

Select: 1-1/2 in. globe valve, parabolic plug: rated $C_v = 28$;* $F_L = 0.9$; $F_d = 0.46$ at rated C_v (see Table 14-2); $C_vF_L/d^2 = 11.2$

$$Re_v = \frac{17300 \cdot 0.46 \cdot 26}{750\sqrt{28 \cdot 0.9}} = 55 \qquad \text{from Equation (5-10),}$$

F_R is about 0.5 for $11.2 = C_vF_L/d^2$ from Figure 5-2 (curve 2)

$$C_{v\ corrected} = \frac{5.1}{0.5} = 10.2 \qquad \text{(OK, 36\% rated } C_v\text{) or choose a reduced port.}$$

The F_R number varies with the internal fluid resistance (K factor) of each valve style (8), i.e., C_v/d_o^2 and the F_L factor (for graphical values, see Figure 5-2).

$$F_R = 0.026\sqrt{Re_v K/F_L} = 0.78\sqrt{Re_v/(C_{vR}/d^2)F_L} \tag{5-11}$$

if flow is completely laminar, or

$$F_R = \sqrt[n]{Re_v/10000} \tag{5-12}$$

where $n = 1 + (890d^4/C_{vR}^2 F_L^{\ 2} + \log(Re_v)$ if flow is transitional.

Since you don't know if the flow is laminar or transitional, calculate F_R with *both* equations, then use the smaller of the two numbers in Equation 5-6 (or get F_R from Figure 5-2). Ignore the viscosity correction completely if Re_v is above 10,000.

---

* Valve size may be increased if calculations show valve is too small.

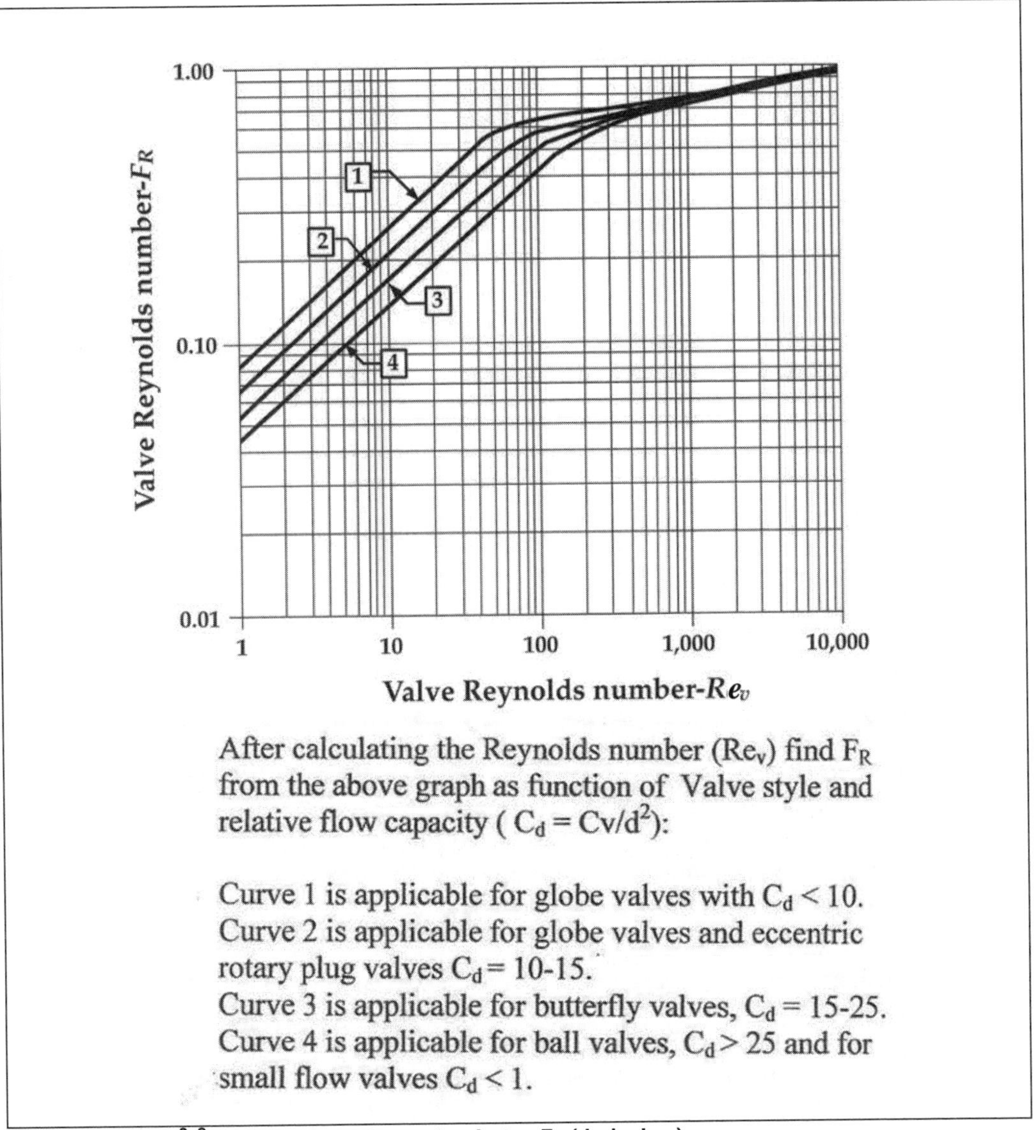

1.00
0.10
0.01
Valve Reynolds number-FR
1
2
3
4
1
10
100
1,000
10,000
Valve Reynolds number-Rev
After calculating the Reynolds number (Rev) find FR from the above graph as function of Valve style and relative flow capacity (Cd = Cv/d2):
Curve 1 is applicable for globe valves with Cd < 10.
Curve 2 is applicable for globe valves and eccentric rotary plug valves Cd = 10-15.
Curve 3 is applicable for butterfly valves, Cd = 15-25.
Curve 4 is applicable for ball valves, Cd > 25 and for small flow valves Cd < 1.

Figure 5-2. Graph[8, 9] of valve Reynolds number factor F_R (d = inches)

For control valves with full-size trim, i.e., where the C_v/d^2 exceeds 8 and further assuming a line size valve and assuming the fluid flow is fully laminar, the C_v can be estimated as follows:

$$C_v = 0.194\left(\frac{qG_f v(F_L)^{2.5}}{\Delta P F_d K}\right)^{2/3} \tag{5-13}$$

For globe valves, you may substitute 0.9 for F_L, 10.0 for K, and 0.46 for F_d. This then makes

$$C_{vGlobe} = 0.059\left(\frac{qG_f\upsilon}{\Delta P}\right)^{2/3}$$

Example: q = 180 gpm; ΔP = 30 psi; υ = 10,000 cst; G_f = 0.6.

$$C_{vGlobe} = 0.059\left(\frac{180 \cdot 0.6 \cdot 10000}{30}\right)^{2/3} = 64$$

This result means you should select a 3 in. globe valve (rated C_v about 105).

For a contoured ball-type control valve with K = 1.4; F_d = 0.8; F_L = 0.6 (From Table 5-4) Equation 5-13 simplifies to:

$$C_{vBall} = 0.077\left(\frac{qG_f\upsilon}{\Delta P}\right)^{2/3}$$

For the above example, this would yield a C_v of 83.7.

Again, a 3 in. size contoured ball valve with a rated C_v of approximately 225 will be appropriate for this job. (Note the uncorrected C_v — assuming turbulent flow is only 25.5!)

If you do not know whether or not the flow is laminar, calculate the required C_v either by the turbulent equation ($C_v = q\sqrt{G_f/\Delta P}$) or by Equation 5-9. Now choose the larger C_v number. Admittedly, this method is not too accurate in the transitional flow area (errors of up to 20% are possible) but is simple and easy to use, at least for an approximation.

Viscosity Conversion

If viscosity is given in centipoises, divide centipoise (μ) by specific gravity (G_f) of liquid to obtain centistokes. If viscosity is given in Saybolt Seconds Universal (t), approximate centistokes by multiplying Saybolt seconds by 0.2 (example: t = 89, v = 18).

Small Flow Valve Sizing

Now for small flow valves ($C_v \leq 0.1$) $F_L \approx 1.0$, we calculate the valve Reynolds number Re_v as before, except we use different F_d factors,[10] namely:

- 0.7 for splined plugs
- 0.3 for short travel diaphragm valves with a flat seating surface
- $F_d = 0.09\sqrt{C_v}/(\text{plug diameter})$, for close clearance tapered plugs

Example: For a 1/8 in. diameter close clearance plug with a $C_v = 0.05$:

$$F_d = 0.09\sqrt{0.05}/0.125 = 0.16$$

For liquids:

$$Re_v = \frac{17300 \cdot F_d \cdot q}{v\sqrt{C_{vR}F_L}} \tag{5-14}$$

For gases:

$$Re_v = \frac{2153 \cdot F_d \cdot q}{v\sqrt{C_{vR}F_L}} \tag{5-15}$$

where q = flow in scfh at 14.7 psia and 60°F.

Now calculate F_R:

If $Re_v \geq 130$,

$$F_R = \sqrt[n]{Re_v/10000} \quad \text{(Note: } F_R \text{ max.} = 1) \tag{5-16}$$

$$\text{where } n = 1 + K + \log(Re_v) \tag{5-17}$$

and since K for small flow valves = 1, $n = 2 + \log(Re_v)$.

For example:

Fluid: CO_2 gas at 85°F (T = 545°R)

$P_1 = 36.3$ psia

$P_2 = 24.7$ psia

v = kinematic viscosity = 8.65 cSt @ 85°F and 14.7 psia (NOTE: Use v at atmospheric pressure if you use q in scfh.)

G_g = specific gravity = 1.516

C_v = valve sizing coefficient (selected) = 0.025

F_L = liquid pressure recovery factor = 1 (from mfr.'s catalog)

F_d = valve style modifier = 0.16 (from mfr.'s catalog)

q = desired max. flow rate = 15 scfh at 85°F and 14.7 psi absolute

$T = 460 + 85 = 545$°R

From Equation 5-15,

$$Re_v = \frac{2153 \cdot 0.16 \cdot 15}{8.65(1 \cdot 0.025)^{0.5}} = 3778$$

and from Equation 5-16,

$$F_R = 5.58\sqrt{\frac{3778}{10000}}\text{; where } n = 2 + \log(3778) = 5.58$$

$$= 0.84$$

Check if the selected valve C_v of 0.025 is sufficient to handle the flow of 15 scfh.

From the turbulent gas equation for sub critical flow* and modified *by* F_R,

$$C_v = q\sqrt{G_g T/\Delta P(P_1 + P_2)}/963F_R$$

$$= 15\sqrt{(1.516 \cdot 545)/(11.6 \cdot 61)}/(963 \cdot 0.84)$$

$$= 0.0193$$

so the selected valve C_v of 0.025 is OK.

Table 5-2. Viscosities and Specific Gravities for Liquids

Liquid	Formula	G_f	T(°F)	v(cSt)
Acetic acid	$HC_2H_3O_2$	1.05	68	1.17
Acetone	C_3H_6O	0.79	68	0.42
Alcohol, ethyl	C_2H_6O	0.79	68	1.52
Alcohol, methyl	CH_4O	0.79	68	0.74
Ammonia	NH_3	0.62	68	0.24
Aniline	C_6H_2N	1.02	60	4.31
Auto oil, SAE 30		0.94	100/210	114/11.2
Benzene	C_6H_6	0.88	68	0.74
Brine, CaCl 25%	CaCl	1.23	68	2.03
Brine, NaC125%	NaCl	1.19	68	1.60
Bromine	Br_2	2.93	60	0.34
Carbon dioxide	CO_2	0.84	-109	0.83
Carbon disulfide	CS_2	1.30	32	0.33
Carbon monoxide	CO	0.84	-314	0.21
Carbon tetrachloride	CCl_4	1.59	77	0.57
Castor oil		0.96	68	1027.00
Chlorine	Cl_2	1.50	-100	0.50

* Remember, there is no sonic velocity if flow is non-turbulent; therefore, critical flow equations don't apply.

Table 5-2. Viscosities and Specific Gravities for Liquids

Citric acid	$C_6H_8O_7$	1.11	68	2.00
Dow Therm A		.79	68	5.60
Dow Therm A		.79	200	1.07
Dow Therm E		.79	68	1.93
Dow Therm E		.79	200	0.53
Ethylene glycol	$(CH_2OH)_2$	1.12	68	16.59
Fluorine	F_2	1.51	-306.4	0.16
Formaldehyde	H_2CO	0.82	60	0.35
Freon 12	CCl_2F_2	1.31	77	0.47
Fuel Oil, #6 (Bunker C)		0.98	85	4080.00
" " " "		0.97	110	1100.00
Gasoline		0.71	60	0.86
Glycerine (100% glycerol)	$C_3H_8O_3$	1.26	68	920.00
Glycol (Ethylene glycol)	$C_2H_6O_2$	1.11	60	17.90
Hydrochloric acid (31.5%)		1.64	60	1.52
Hydrogen	H_2	0.074	-422	0.011
Hydrogen chloride	HCL	0.91	68	2.67
Hydrogen fluoride	HF	0.97	68	0.13
Hydrogen sulfide	H_2S	0.79	60	
Isopropyl alcohol	C_3H_8O	0.79	68	2.94
Linseed oil		0.93	60	55.90
Kerosene		0.82	85	2.99
Mercury	Hg	13.57	60	0.12
Napthalene	$C_{10}H_8$	1.15	68	398.00
Nitric acid (40%)	HNO_3	1.25	68	1.25
Nitrogen	N_2	0.85	-320	0.19
Oxygen	O_2	1.20	-297	0.17
Phosphoric acid	H_3PO_4	1.15	68	1.85
Propylene glycol (60%)	$C_3H_8O_2$	1.04	68	8.99
Sodium hydroxide	NaOH	1.27	68	6.20
Starch (4% Solution)	$(C_6H_{10}O_5)x$	1.02	100	100.00
Sulfuric acid (100%)	H^2SO_4	1.83	68	13.91
Toluene	C_7H_8	0.87	68	0.68
Trichloroethylene	C_2HCl_3	1.46	80	0.39
Turpentine		0.87	60	1.71
Water, fresh	H_2O	1.000	60	1.00
Water,sea (4%saltbywgt)		1.03	60	1.05

Table 5-3. Viscosities and Specific Gravities for Gases at Atmospheric Pressures

Gas	Formula	G_g	T(°F)	v(cSt)
Acetic acid	$HC_2H_3O_2$	2.07	246	5.75
Acetone	C_3H_6O	2.01	212	4.91
Acetylene	C_2H_2	0.907	32	7.99
Air	N_20_2	1.000	64	15.10
Alcohol				
Butanol, butyl	$C_4H_{10}O$	2.56	242	10.94
Ethanol, ethyl	C_2H_6O	1.59	212	7.20
Methanol, methyl	CH_4O	1.11	152	11.77
Propanol, propyl	C_3H_8	2.07	212	4.76
Ammonia	NH_3	0.596	68	13.71
Argon	A	1.379	68	13.38
Benzene	C_6H_6	2.70	58	2.23
Bromine	Br_2	5.52	55	2.21
Carbon dioxide	CO_2	1.5290	68	8.06
Carbon disulfide	CS_2	2.63	68	2.88
Carbon monoxide	CO	0.97	59	14.51
Carbon tetrachloride	CCl_4	5.31	170	3.65
Chlorine	Cl_2	2.486	68	4.44
Cyanogen	C_2N_2	1.806	63	4.51
Ethane	C_2H_6	1.049	63	7.08
Ether	$(C_2H_5)_2O$	2.56	58	2.28
Ethyl chloride	C_2H5CL	2.220	32	3.27
Ethylene	C_2H_4	0.975	68	8.60
Fluorine	F_2	1.31	68	13.98
Freon II	CCl_3F	5.04	68	1.77
Freon 12	CCl_2F_2	4.2	68	2.46
Helium	He	0.138	-431	0.89
Helium	He	0.138	68	117.06
Hydrogen	H_2	0.0695	-431	0.37
Hydrogen	H_2	0.0695	68	104.90
Hydrogen bromide	HBr	2.819	66	5.35
Hydrogen chloride	HCl	1.268	64	9.32
Hydrogen cyanide	HCN	0.93	68	8.95
Iodine	I_2	8.76	338	2.93
Isopropyl alcohol	C_3H_8O	2.08	212	5.55
Krypton	Kr	2.89	59	6.96
Mercury	Hg	6.93	523	11.05

Table 5-3. Viscosities and Specific Gravities for Gases at Atmospheric Pressures

Methane	CH_4	0.554	68	15.41
Methyl acetate	$C_3H_6O_2$	2.56	212	4.14
Methyl chloride	CH_3Cl	1.74	59	4.89
Methyl iodide	CH_3I	4.47	811	14.66
Neon	Ne	0.69	68	37.53
Nitric oxide	NO	1.04	68	15.01
Nitrogen	N_2	0.97	81	15.66
Nitrosyl chloride	ClNO	2.26	59	4.12
Nitrous oxide	N_2O	1.52	80	8.33
Oxygen	O_2	1.10	66	15.23
Pentane (n)	C_5H_{12}	2.49	77	2.30
Propane	C_3H_8	1.52	64	4.32
Propylene (propene)	$C_2H_4(CH_2)$	1.45	62	4.71
Sulfure dioxide	SO_2	2.21	69	4.73
Saturated steam (30 psia)	H2O	0.62	250	12.20
Toluene	C_7H_8	3.18	68	1.83
Xenon	Xe	4.53	68	4.15

NOTE: For pressures other than atmospheric, multiply listed v by $14.7/P_1$, where P_1 = actual inlet pressure in psia.

SIZING CONTROL VALVES FOR VISCOUS FLUIDS - AN EXPLANATION

I always wondered how a ball valve with hardly any obstruction can pass the same amount of tar* for a given pressure drop and flow coefficient (C_v) than a cage valve with, say, 100 small, drilled holes even though it has the same C_v as the ball valve. Instinct tells us this could not be so despite the fact the previous ISA standard told us so. Not believing it, I decided to do something about it and came up with the following revised method.

Also, the F_R for small flow valves derived from my new equations did not agree with values given in Figure 1 of the former ISA-75.01.01, Flow Equations for Sizing Control Valves (now revised), and yield more accurate results typically increasing the required C_v for small flow valves by a factor of four or more over the current method in Reynolds numbers below 100. The former ISA (single) F_R curve was based on a limited number of tests with full-sized globe valves while the new values are based on air and water tests with small flow valves, or rotary valve types in addition to globe valves.

From data published in Perry's *Chemical Engineers Handbook*, Sixth Edition, Table 5-15, it appears that more streamlined valves (e.g., angle valves and gate

* Ignoring the F_L factor, which affects the Reynolds number.

valves) exhibit F_R factors close to Figure 5-2 curves, while disk and plug-style globe valves (due to their more abrupt flow pattern) tend to favor the former ISA F_R values when corrected with the proper F_d values.

From this data, it appears that the "pseudo C_v", i.e., the "laminar flow C_v"or $C_vF_RF_L$, is identical for all valve types with full size trim and at low Reynolds numbers ($Re_v < 100$).

From this, it can be stated that the required C_v (valve capacity needed to pass viscous flow for a given Reynolds number) is

$$C_{vReq} = A_p/(0.026\sqrt{K})$$

where 0.026 is the area in square inch per 1 C_v when the head loss coefficient $K = 1$. A_p is the cross-sectional area of the given pipe size in inch^2 (same as valve size!).

Furthermore, from the ISA equation, $C_{vReq} = C_i/F_R$, where C_i is the C_v calculated assuming the flow pattern is completely turbulent. This makes $C_i = A_p\sqrt{Re_v}/F_L$, and since the valve Reynolds Number for liquids is,

$$Re_v = \frac{17300 \cdot F_d \cdot q}{v\sqrt{F_L C_{vReq}}}$$

$$C_{vReq} = 0.194\left(\frac{qG_f v(F_L)^{2.5}}{\Delta P F_d K}\right)^{2/3}$$

This is Equation 5-13.

Since F_L is approximately proportional to the relative valve capacity (C_v/d^2), we may, for estimating purposes only, substitute

$$F_L \cong 0.96 - [0.0025(C_v/d^2)^{1.5}] \text{ in Equation 5-13.}$$

Furthermore, F_d is somewhat related to F_L (for example a large valve, such as a ball valve, has a low F_L but a high F_d). We can thus further estimate that

$$F_d = 0.4/F_L \text{ and, finally, } K = 890\, d^4 / C_v^2.$$

This then makes Equation 5-13

$$C_v = 0.00386\left\{\frac{qG_f v\left[0.96 - 0.0025\left(\frac{C_v}{d^2}\right)^{1.5}\right]^{3.5} C_v^2}{\Delta P d^4}\right\}^{2/3}$$

NOTE: The selected C_v should not be less than 10 d^2 and no more than *30* d^2.

I suggest you input the customer's pipe size for *d*. The above equation will give you a ball park C_v number which can aid in selecting a valve type. Using the actual valve coefficients (rated C_v, F_L, F_d), recalculate the required C_v from Equation 5-9.

METRIC UNITS

Normally, it does not bother you that the rest of the world thinks in metric terms. However, an occasion may arise when you will have to do a project for an overseas plant, and you will be confronted with liquid flow rates of cubic meters per hour instead of gpm. Luckily, the C_v numbers work equally as well in Europe. For example:

$$C_v = 1.17q/\sqrt{\Delta P/G_f} \qquad (5\text{-}18)$$

where:

q = m^3/hr of liquid

ΔP = bar,

G_f = specific gravity (water @ 60°F = 1)

As a matter of fact, in Germany, they go so far as to eliminate the 1.17 constant and then call the flow coefficient K_v; hence, $K_v = 0.85\ C_v$, or $C_v = 1.17\ K_v$.

Complications arise only if you are confronted by newtons per square meter, also called pascal or Pa. To get back to the more familiar bar, divide Pa by 100,000 or one bar = 100,000 Pa, and one bar = 14.5 psi. One kilopascal (kPa) = 0.145 psi. There is also the less often used kg/cm2, which is 0.98 bar or 14.2 psi.

Remember also, that degrees Celsius is used abroad where °F = (°C x 9/5) + 32, and in absolute terms, °K = 273 + °C. Converting flowing units:

1 m^3/hr liquid = 4.4 gpm

1 m^3/hr gas = 35.3 ft^3/hr

kg/hr = 2.2 lbs/hr

You are now fully equipped to work abroad.

WHAT SIZE VALVE TO CHOOSE

After you have labored hard to select the final required valve C_v suitable for maximum flow requirements, and having considered the **F** effects (F_L, F_p, F_R) you are now confronted with the final choice—the valve size.

If the pressure drop is low, say, 5 to 15 psi, it is always safe to start with a line size valve. Keep in mind that the person in the Piping Design Department already sized the downstream or upstream pipe to pass the design flow rate at a reasonable velocity. Incidentally, valve outlet velocity should not bother you unless there could be a noise problem with high pressure gas or steam reduction (see aerodynamic noise discussion in Chapter 14), or where you have flashing liquids (if the downstream pipe is too long, your valve will cavitate because of downstream pressure buildup caused by high mixed-phase velocity). This then leaves only the C_v rating of the valve. In case you have not yet selected a certain brand because you are partial to the color of the valve, i.e., green or red, here is a hint:

Table 5-4. Typical Flow Coefficients at Rated Travel

Type	C_v/d^2	F_L	C_vF_L/d^2	C_vF_p/d^2		for	
				D/d=1.5		D/d=2	
Single-seated globe valve							
Parabolic plug, flow-to-open	10	0.9	9	9.8	(4.3)	9.6	(2.4)
Double-seated globe valve	12.5	0.9	11.3	12	(5.3)	11.7	(2.9)
High capacity cage valve							
to size 3 in. only	14	0.9	12.5	13.5	(6.0)	12.9	(3.2)
Eccentric rotary plug valve	12 0.	85	10	11.6	(5.1)	11.3	(2.8)
Butterfly valve 60° open	17	0.68	11.5	15.8	(7.0)	15	(3.8)
Butterfly valve 70° open	27	0.57	15.5	23	(10.2)	20.8	(5.2)
Fluted butterfly valve @ 70°	25	0.7	17.5	21.8	(9.7)	19.8	(5.0)
Contoured ball valve @ 90°	25	0.6	15	21.8	(9.7)	19.8	(5.0)
Angle, venturi type,							
flow-to-close	22	0.5	11	19.8	(8.8)	18.3	(4.6)

D = inches

The C_v values are per *valve* diameter (inch) squared. The C_v values in parentheses are per *pipe* diameter (inch) squared. For example: A 6 in. eccentric rotary plug valve has a wide-open, C_v flow coefficient of 6^2 x 12 = 432. However, when installed between reducers in a 12 in. pipe, the available flow coefficient is only 6^2 x 11.3 = 407 (see column C_v/d^2 for D/d = 2), or 12^2 x 2.8 = 403 using the figures in parentheses.

The above values are conservative and vary somewhat from size to size. You may notice that when high-capacity valves are used under choked flow conditions (such as steam pressure reduction), they all tend to lose their high C_v

advantage, as you can see from the $C_v\ F_L/\ d^2$ column! Explain this to your valve vendor.

Starting out with a line size valve, divide your desired maximum C_v value by 0.8. First, this will give you a 10% safety factor and second, it will guard against the ±10% tolerance in the C_v rating of the valve given by the manufacturer. Now divide this new C_v number by the pipe size squared. For example: From the flow data, you require a maximum $C_v = 123$. Your line size is 4 inches. First divide 123 by 0.8 to obtain 154. Now divide this by $(4\ \text{in.})^2$. $154/4^2 = 9.6\ C_v/d^2$.

From the data given in Table 5-4, you have the following choices:

- A 4 in. single-seated globe valve, $C_v/d^2 = 10$; C_v rated $= 10 \times 4^2 = 160$.
- A butterfly valve at 70° opening (if other considerations such as leakage, noise, etc. permit). Here a 3 in. butterfly valve installed in a 4 in. pipe has a flow capacity higher than $10.2 \times D^2$ (see figures in parentheses), i.e., more than $4^2 \times 10.2 = 163$ C_v.

Having thus determined the size (and type) of valve, you now should verify that you can meet the minimum required flow coefficient that your process requires. Here you have to go to the manufacturer's C_v tables (usually printed in increments of 10% travel). Look up the approximate valve travel when the valve is operating at your minimum C_v. This travel should not be less than 5% of the total valve travel or less than the percentage of the valve dead band (with tight packing).* This is usually 5% with TFE packing to 15% for laminated graphite packing if *no* positioner is used. For a typical TFE-packed valve and no positioner, this limits the valve travel to 10% from the shut-off position. The only exceptions are certain soft seat plugs that "dip" through their seal ring, ball valves, elastomer-lined butterfly valves with angle-seated vanes, or certain diaphragm or pinch valves where there is no "hard" seating action. If a standard valve forces you to throttle below 5% travel, then your only choice is to use a larger and a smaller valve in parallel and split-range the actuator signal, or better yet, use two separate signals from your computer so that the small valve will open first, and then the larger valve starts when the small valve is about 80% open.

In summary:

- Choose a valve that has a rated C_v at least 20% larger than your maximum requirements.
- See that the minimum required C_v occurs at least above 5% of the valve travel (check vendor's catalog).

---

* You may have to use 10% if the 5% travel C_v is not known.

ADJUSTABLE TRAVEL - ADJUSTABLE C_V

As mentioned previously, one of the vexing problems confronting the instrument engineer is the fact that the final required flow capacity that the selected valve needs is sometimes less than half of what was specified. The result is an effective signal span of less than 40% (4 - 10.4 mA) in case of a linear plug characteristic and less than 75% (4 - 16 mA) with an equal percentage flow characteristic.

Normally, the controller or the computer does not have a problem controlling the loop with a reduced signal span since the signal resolution is still about one in one thousand for 50% of signal span. However, the minimum controllable flow at 5% travel is higher (as a percentage of the maximum flow), and, worse, the control valve dead band is more than twice as much as a percentage of the usable signal span, thereby affecting loop stability!

It would, therefore, be nice if we could reduce the rated valve C_v after startup, hopefully with the valve staying in the line. This is preferable to sending the valve back to the maintenance shed for installation of a reduced trim.

Well, valves are now available where you can adjust the rated C_v within reasonable limits, and it may be useful to get familiarized with them. The illustration in Figure 5-3 shows such a valve in a schematic way to demonstrate the principle of adjustment in flow capacity. Here, a conventional spring-diaphragm actuator (1) moves a linkage (2) up and down, which, in turn, connects to a valve stem (3). The stem, in turn, moves a valve plug (4) against a seat ring (5). In the normal actuator position, A, the valve plug travels the maximum distance, H, which in turn yields the maximum valve C_v, say 300.

With an equal percentage plug contour, you get a typical "inherent flow characteristic" as shown in Figure 5-3(B). Now assume that after installation we find the valve needs a flow coefficient of only 105 (equivalent to 70% of the present travel). To increase the usable signal span (= actual travel) to, say, 90% from the actual 70% (allowing a 10% safety margin), we can now relocate the actuator (1) along mounting bracket (6) to a new position, B.

Owing to a greater distance L_2 (not to scale) from the pivot point (7), the excursion of the lever (2) is reduced, resulting in a new travel H of 78% of the travel we got at actuator position A. Assuming H was 2 in. at position A, we now get 1.56 in. at B. The equal percentage characteristic at this point yields a C_v *of* 150, as you can see from Figure 5-3(B). With the reduced travel, we can now "spread" the curve over the available 100% signal span or over 1.56 in. plug travel. The result is shown in Figure 5-3(C).

To help you understand how this graph was developed, assume you stroke the plug to 60% travel in actuator position B. Reducing the travel now by 22% (to 78% of 2 in., or 1.56 in. travel) will yield a new plug percentage travel of 0.78 x 0.6 = 0.468 or 46.8%. The actual C_v of the plug at this point is about 37 (from Figure 5-3(B)). However, since 46.8% of the actual plug travel corresponds to 60% of the actuator travel in position B, you have to plot the 37 C_v at the 60% travel position (equivalent to 60% actuator travel, or 60% of signal) in order to construct the curve in Figure 5-3(C).

It is interesting to note that when comparing the two curves at 10% travel, we find a C_v value of 10 for the full travel plug but only 7.5 C_v for the reduced travel

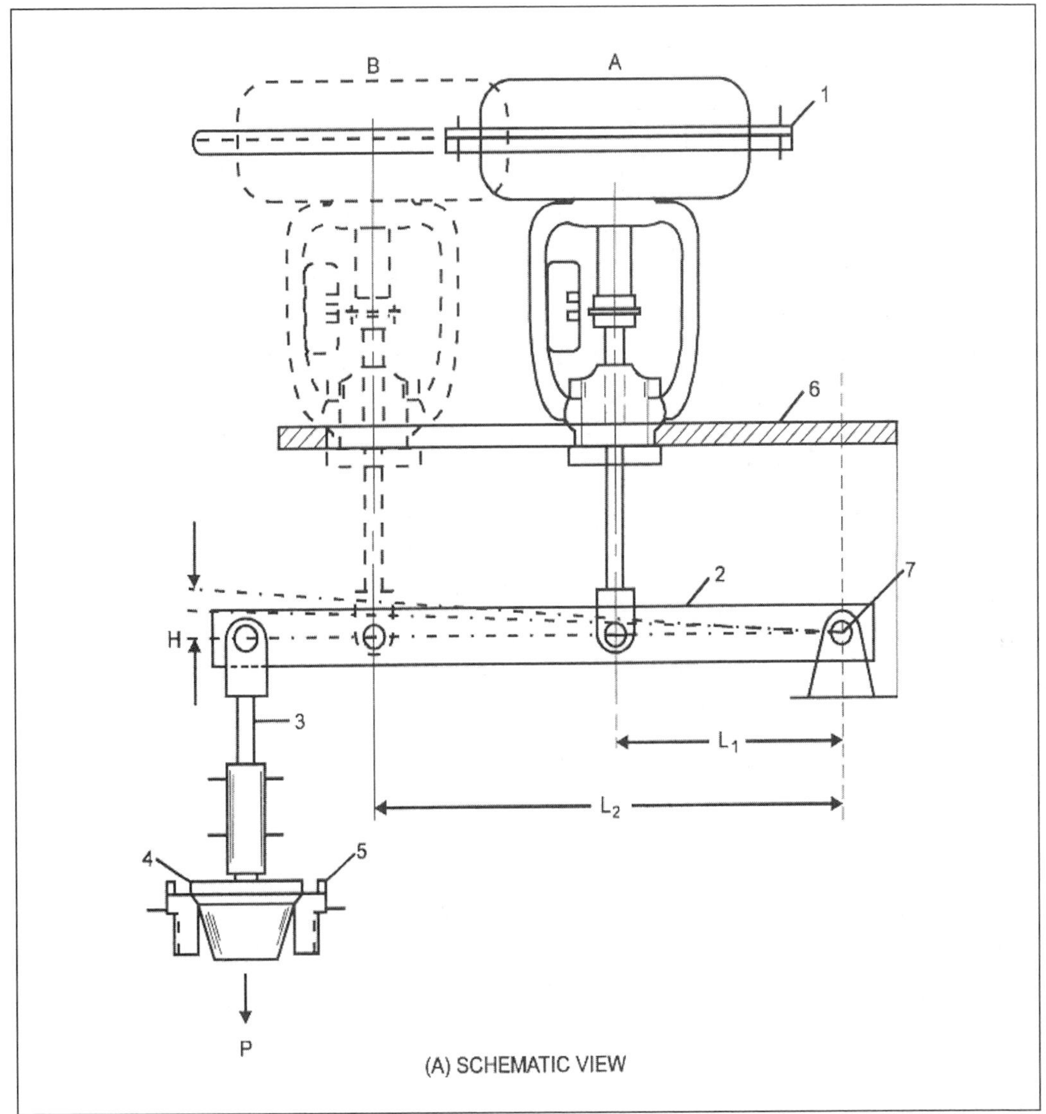

Figure 5-3. Control valve with equal percentage characterized plug and adjustable travel.

plug, yielding a somewhat wider, useful rangeability, in this case, (105 C_v / 10 C_v = 10.5:1 over 105 C_v /7.5 C_v = 14:1!).

Of greater significance is the fact that by locating the actuator (1) closer to the valve stem (3), we greatly increased the available plug force P, in the present example by (1/0.78) = 1.28 times. This not only complements the typical increase in pressure drop across the valve with decreasing flow capacity but also reduces the effects of packing box friction on actuator positioning accuracy (dead band). As a matter of fact, you get a double effect: you get an increase in actuator force over the constant packing friction, but you also divide this friction over the larger signal span!

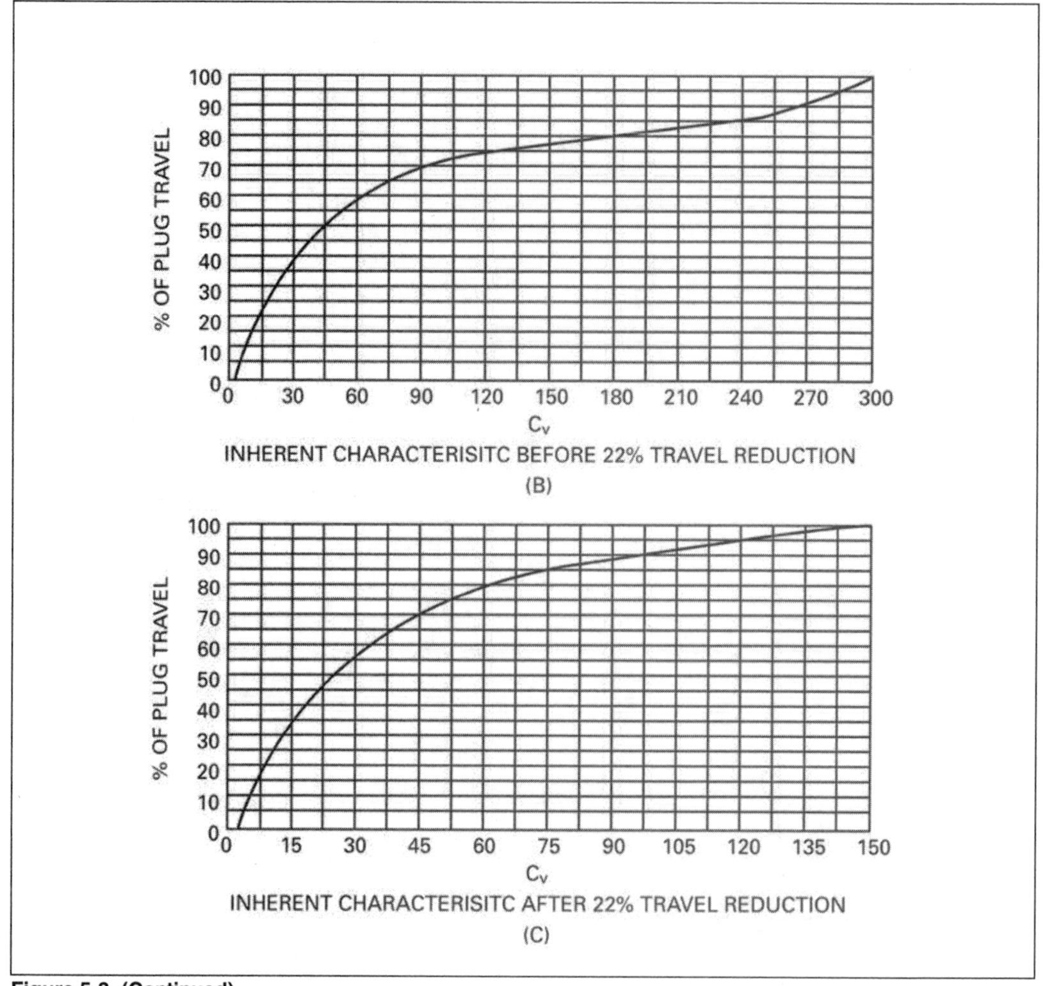

Figure 5-3. (Continued)

Again, using the above example: (a) friction, as seen by the actuator, is reduced to 78%, and (b) lower friction is distributed over 100/78% signal span. Total dead band (as caused by valve friction only) reduction is therefore 39%, [100 x (1 – 0.78^2].

While the above scheme has beneficial effects concerning actuator forces and dead band, we should not expect improvements in the "installed" valve gain. Granted, the gain change (see Figure 8-1) is now spread over a greater signal range; however, the percent of flow at which the "limit in gain change" is exceeded is still about the same.

The above example is conservative since it assumes an equal percentage plug characteristic. If the case had been a linear valve, then with the same C_v decrease, the plug travel would have to be reduced to 40% of the original travel, yielding an actuator force increase of 250% and a packing friction reduction of 84%! Simply reducing the actuator travel of a conventional valve by merely recalibrating the valve positioner will not do the same job. You still get the same actuator force and the same dead band, except the dead band is now divided into a larger signal span, thereby resulting in a lower percentage of signal offset!

The travel of a valve plug or rotary closure member cannot be reduced indiscriminately. For example, with some rotary valves, the seating friction near the point of shut-off becomes very high. Also, most parabolic plugs don't extend their contoured portion all the way to the seating bevel, but are cylindrical for measuring purposes. Therefore, a reduction of valve travel to less than 40% of the original travel does not seem advisable. Also, such travel adjustment is not practical with valves that have an inherent low rangeability, i.e., a low (<20:1) ratio between maximum C_v and minimum C_v.

REFERENCES

1. Baumann, H. D., "The Introduction of a Critical Flow Factor for Valve Sizing," *ISA Transactions,* 2:107-111 (1963).
2. ANSI/ISA-75.01.01 (IEC 60534-2-1 Mod)-2007 Flow Equations for Sizing Control Valves. Research Triangle Park, NC: ISA.
3. Driskell, L. R., "New Approach to Control Valve Sizing," *Hydrocarbon Processing,* July: 111-114 (1969).
4. Bulletin No. SM-6, "Control Valve Sizing." Portsmouth, NH: H. D. Baumann Assoc., Ltd., (1991).
5. Baumann, H. D., "A Case for Butterfly Valves in Throttling Applications," *Instruments & Control Systems,* May (1979).
6. Baumann, H. D., "Effect of Pipe Reducers on Control Valve Capacity," *Instrument & Control Systems,* pp 99-102, Dec. (1968).
7. McCutcheon, E. B., "A Reynolds Number for Control Valves," Symposium on Flow, its Measurement and Control in Science and Industry, Vol. 1, Part 3, 1974.
8. Baumann, H. D., "A Unifying Method for Sizing Throttling Valves Under Laminar or Transitional Flow Conditions," *Transactions of the ASME, Journal of Fluids Engineering,* pp 166-169, March (1993).
9. George, J. A., "Evolution and status of non-turbulent flow sizing for control valves", Monograph, ISA 2001.
10. Baumann, H. D., "Viscosity Flow Correction for Small Control Valve Trim." ISA PAPER 90-618, Research Triangle Park, NC: ISA, October (1990).

6

SIZING AND SELECTION—LET THE COMPUTER DO IT ALL!

There are basically two types of sizing software programs: The first lets you pick the valve type then gives you the specifics selected from the database, such as rated C_v, F_L, and F_d values, allowing you to do the right sizing calculations. The second type lets you enter only the flow conditions, and then the computer calculates the C_v and selects the right valve, usually the best economical choice.

There are problems with both methods. In the first case, you are forced to select a valve type and vendor before you have a chance to get a competitor's bid or evaluate a more economical valve type. In the second case, your sizing program has to start with some "assumed" parameters such as F_L and F_d values. The computer then chooses the best valve type and size to fit this application. However, the chosen F_L or F_d may be different from the assumed one and, as a result, you now have to repeat the sizing calculations. Not quite an ideal situation. Nevertheless, computers can be great time savers especially when it comes to calculating the effects of reducer losses or viscosity. Both of these require otherwise time-consuming iterations (repeat calculations in which you have to assume a valve size or C_v value, do your calculation, find out the assumed valve size was too small, pick a larger valve size or C_v value and then start over, usually several times!).

Basic computer sizing software can calculate the required C_v value usually for maximum, normal, and minimum flow conditions. The software tells you if the valve will cavitate or flash and can calculate the aerodynamic or hydrodynamic valve noise.

Don't be impressed by the computer's ability to calculate a C_v number to the sixth decimal place. This is meaningless. Correct results depend on an error free program and selecting the right input data. I would always recommend using a slide rule or a simple hand-held calculator to double check the computer's result, especially if you use a brand new software.

Better programs will go further and select the right valve size and type for your flow condition based on such required parameters as end connections (flanged, welded, screwed, etc.); leakage requirements (Class I, II, IV, V, or VI);

should the valve be corrosion resistant (no, mild, very); the flow characteristic (equal %, linear); and perhaps others. Based on the above, the computer will search the database and select the most suitable and economical choice. Some will even calculate the required actuator size and bench range. However, do not expect the program from Company "A" to select a valve from Company "B." After all, there are limits to benevolence.

More advanced programs go a step further and calculate the pressure drop in your piping system (you have to provide the input) and calculate the "installed" flow characteristic of the valve and the "gain" (d_q/d_s) of the valve under various flow conditions. This way you can see if the control loop will be stable[1].

When selecting a program (most vendors will give one to you for free), check the following:

- Valve sizing should be per current ANSI/ISA-75.01.01 standard or the corresponding IEC standards.
- Noise equations should follow IEC Standard 60534-8-3 (for gas) or IEC 60534.8.4 (for liquids).
- The required maximum C_v should not be more than 85% of the rated C_v of the selected valve.
- The minimum required C_v should be greater than the C_v of the selected valve listed at 5% of valve travel (if a positioner is used), or 10% if the packing friction may be high and no positioner is used.
- Don't get scared by the ominous warning "choked flow." This has no meaning other than it requires different sizing equations.
- Ignore calculations of inlet and outlet velocities. These mean nothing except on aerodynamic noise where you should avoid fluid velocities in valve outlet above 0.3 Mach!
- Pay attention to "cavitation" or "flashing" warnings. They may spell trouble (see Chapter 14).

---

1 Some people have the misconception, that the gain of the control valve has to be "one" +/–. This is not necessary and typically not possible (since you never use the full signal span). What is important is, that the gain (whatever it is) stays reasonably constant with change of flow conditions see Chapter 8.

PARTIAL TABULATION OF VENDORS OFFERING COMPUTER PROGRAMS FOR CONTROL VALVE SIZING AND SELECTION

Engineered Software, Inc.
P.O. Box 2514
Olympia, WA 98507

Fisher Controls International, Inc.
205 S. Center St.
Marshalltown, IA 50158

Gulf Publishing Co.
Houston, TX

ISA
P.O. Box 12277
Research Triangle Park, NC 27709

Instrumentation Software, Inc.
P.O. Box 776
Waretown, NJ 08758

Masoneilan Dresser
Dresser Valve & Controls Division
275 Turnpike Street
Canton, MA 02021

MESA, Inc.
P.O. Box 15004
Worcester, MA 01615

Flowserve Incorporated
PO Box 2200
Springville, UT 84663

7

WHAT ABOUT FAIL-SAFE?

One of the important benefits of spring-diaphragm-actuated control valves is their ability to "fail-safe" by the action of the pre-compressed actuator spring, i.e., to close or open the valve on air failure or, more typically, breakage of the signal wire or controller malfunction.

Cylinder-actuated valves do not fail safe without the additions of springs or stored air pressure in a separated cylinder kept there by a pressure-actuated lockup valve. Sometimes, the flow direction of a single-seated globe valve is reversed so that the fluid forces help to close the valve plug (if fail-close is desired).

Typically, electric actuators do not provide fail-safe action unless a standby power source, such as a relay switch-actuated battery pack, is provided. Here, of course, periodic maintenance and recharging of the battery is a must.

The fail-safe direction (open or closed) is usually dictated by safety concerns. Typically,[1] "fail-close" is specified if the fluid to be handled is flammable (such as natural gas going to a burner) or is a heating medium going to a heat exchanger where adding too much heat to a kettle can cause boiling or can cause an exothermic reaction of a chemical. Most valves handling chemicals are specified as "fail-close" valves.

However, in some heating applications, fail-open action is more advantageous. As an example, consider the steam tracing of a crude oil pipe. Here, too little viscosity (too much heat) is probably better than having the fluid come to a near standstill, particularly in the middle of winter. Another fail-open application is controlling cooling fluid to a motor bearing where the worst effect may be a decrease in efficiency.

Ask yourself the question, "What would happen if this valve suddenly opens (or closes) upon failure of the air compressor or upon signal failure (computer breakdown, wire breakage, lightning strike)?" When in doubt, check with your process people. Another question to keep in mind is; "does the downstream safety valve have sufficient flow capacity to handle the flow of a pressure reduc-

1 In 75% to 80% of all applications.

ing valve when it fails wide open?" In the same vein, consider the effects on a trapped liquid experiencing thermal expansion when a valve that is supplying a heating medium fails in the closed position.

In some cases you may wish the valve would remain as is in the last travel position, particularly if you are dealing with a short power failure to the control system. Here, the standard electric gear-type actuator is ideal. Spring-diaphragm and other pneumatic actuators have to use lockup valves to trap air pressure in the actuator if the fail-in-place mode is desired.

To guard against compressed air failure, have the pneumatic signal to the actuator pass through a valve that is kept open by the pressure in a separate air supply line. As soon as this supply pressure diminishes, it will close the lockup valve and trap the actuator pressure. To guard against signal failure, here a two-way solenoid valve can take the place of the pneumatic lockup valve and trap the air upon electrical power failure.

For added safety, you may want to consider the flow direction through the valve to guard against the consequences of valve stem failure. Here, for example, you want to specify the direction of fluid flow as "flow-to-open" in case the valve should fail open, or "flow-to-close" if closure is desired.

However, specifying a "flow-to-close" plug can be tricky since it can lead to instability (plug may slam closed), see Chapter 10.

When considering the safety aspects of your control valve application, evaluate the following possible consequences in order of increasing severity:

- decrease in product yield,
- shut down of process,
- damage to process equipment, and
- injury to personnel or loss of life.

8

WHY MOST PEOPLE CHOOSE "EQUAL PERCENTAGE" AS A FLOW CHARACTERISTIC

One of the most vexing questions asked is, "What inherent flow characteristic should my valve have?" First, a definition: The "inherent flow characteristic" is the relationship between the flow coefficient C_v and valve travel. If it is linear, then you get (theoretically) the same amount of C_v increase for an equal increase in travel. Typical values are a 9.8% increase in C_v for every 10% increase in travel. Aha! You wonder why not 10% C_v? Well, at say, 1% travel, the flow is already 2% of maximum C_v due to clearance flow between the plug and seat ring. Hence, the change is only in 9.8% per 10% increments (see Figure 8-1).

For "equal percentage characteristic," the typical C_v increase is about 43% above the previous number for each 10% of travel; for example, at 20% travel the C_v number is 143% of that at 10% travel; assuming the C_v at 40% travel is 10; then at 50% travel it is 14.3; and at 60% travel, 20.5 and so on. This means that the flow increases exponentially instead of linearly. This characteristic is typically presented by a straight line on semi-logarithmic paper.

Finally, there is the "quick-open" characteristic, which means the valve plug is no longer contoured but is usually a flat disk. Here, the C_v increase is approximately linear, up to a travel equivalent to 1/4 of the orifice diameter. After that, the flow stops increasing for all practical purposes. You may safely ignore this characteristic for throttling control valves.

Now a word of caution: After the valve is installed into a piping system, the above flow characteristics are only applicable when:

- the pressure drop remains constant with flow changes;
- there will be no onset of flashing, cavitation, or sonic flow at certain flow conditions (quite possibly due to changes in the F_L factor of the valve, with changes in valve travel); and
- the valve is not installed between reducers.

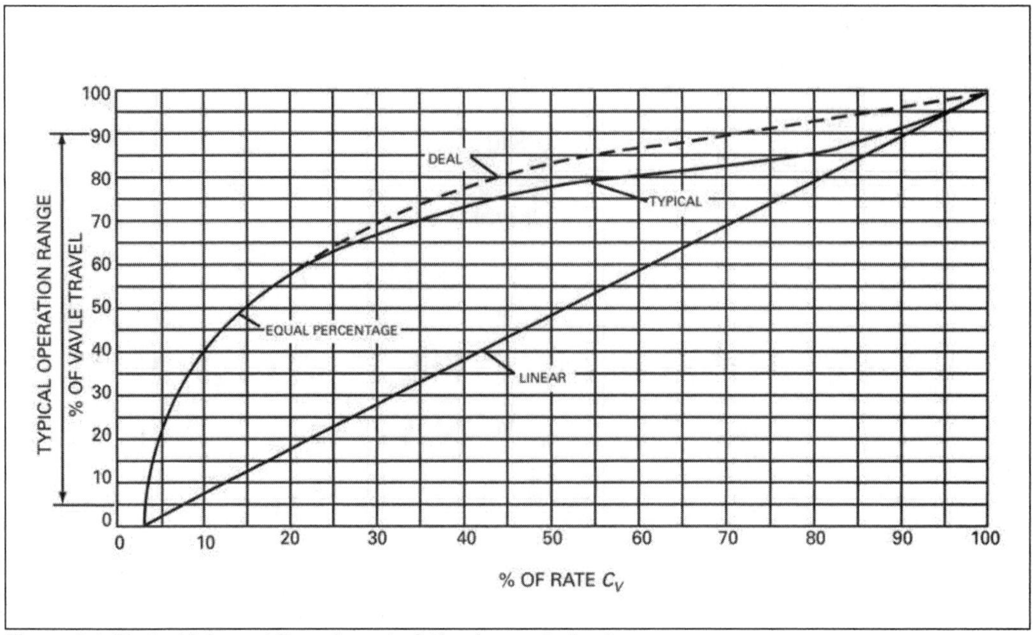

Figure 8-1. Typical inherent flow characteristics for control valves.

As you can readily see from the above, ideal conditions that prevent any of the above from occurring will prevail only in the manufacturer's flow lab. So, to account for the real world, we have to deal with the "installed flow characteristic," which is the relationship of flowing quantity to valve travel, accounting for the difference in pressure drop between maximum and minimum flow; the possibility of choked flow along the line; and the distortion that the reducer head losses create in the inherent valve flow characteristic. This installed characteristic is what finally affects your controller stability.

An ideal Controller PID (proportional, integral, and derivative action) setting will demand a constant "gain" of the final-control element. This gain is the rate of change in flow *(q)* for a given change of signal. Usually, the valve travel *(h)* is proportional to the signal (s). This makes $\delta(q)/\delta(s)$ = constant, for the ideal valve.

Ideally then, what we want is a "linear" installed flow characteristic which is about a 10% increase in flow (not C_v) for every 10% increase in travel (this would result in an installed "gain" of the valve of "one"). However, nothing in life is ideal, and we, therefore, have to compromise. As a rule of thumb, the valve gain can vary ±50% from the average constant gain value (does not need to be 1.0) without requiring an adjustment in the controller's PID settings. This means, for example, that one can accept an installed characteristic where the flow can vary 5% per 10% travel at one end and 15% per 10% travel on the other. This certainly makes life easier. Figure 8-2 shows an example of a typical installed valve flow characteristic. Here, the controller was tuned at 20% flow or 28% of signal.

The slope of the curve at that point (15% flow per 20% signal) serves as a reference. In this case, it just so happens that the gain changes in either direction. Therefore, the ± 50% (i.e., 1.5 and 0.5) "constant gain" slopes determine the limits

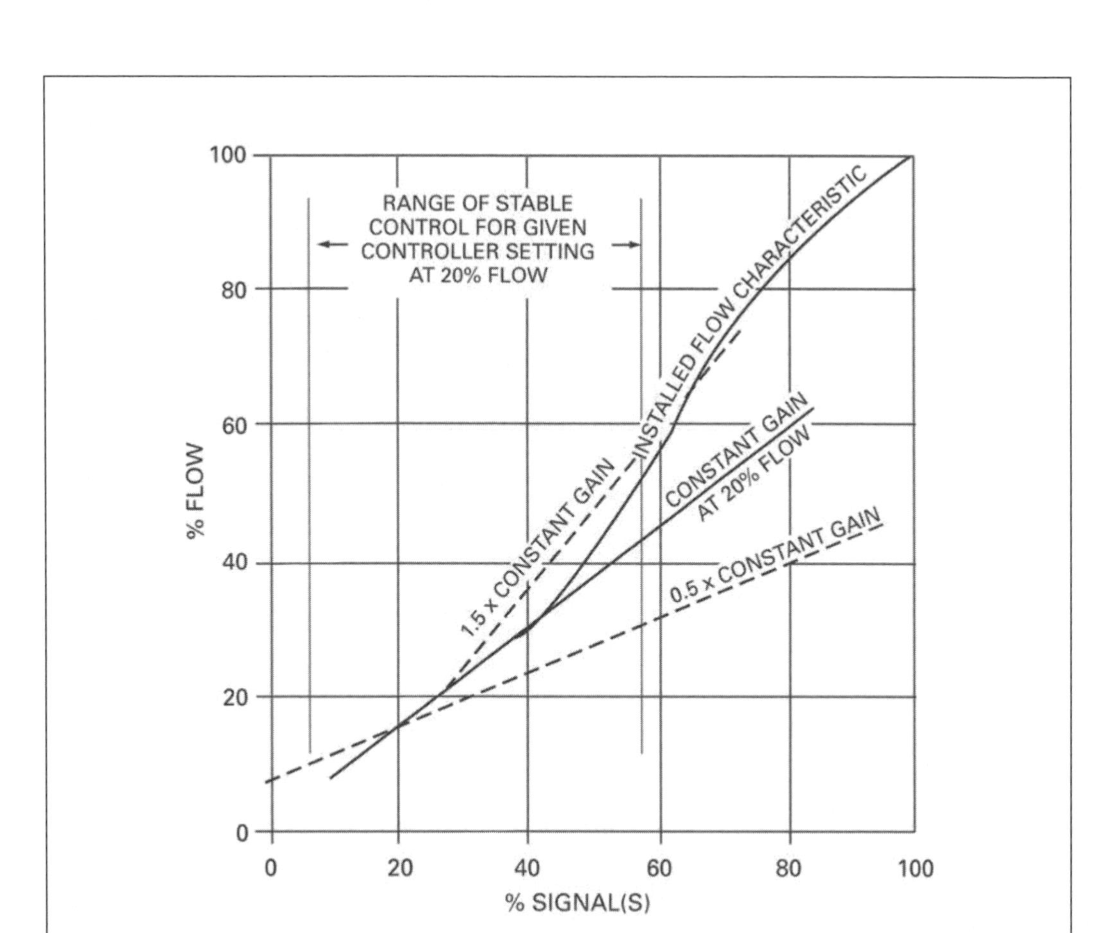

Figure 8-2. Typical installed flow characteristic of a control valve having a "constant" gain of 0.75, with super-imposed 50% control valve gain variations from original controller setting, indicating limits of expected stable controller performance. Retuning of controller may be necessary if flow range exceeds those shown limits.

of controller stability, which are 10% flow (8% signal) and 57% flow (58% signal). Based on this information, we can now make the following rules.

First, check your process and find out what the ΔP across the valve is at the minimum required flow rate and what ΔP is at the maximum flow rate. Establish a ratio between the two.

Second, if the ratio is less than 2:1, choose a linear inherent valve characteristic. If the ratio is above 2:1, choose an equal percentage plug.

* Change the proportional band of the controller to 50%, if the average gain is around 2 for example.

So much for science. Now let us look at the more mundane aspects of the subject. As I have already stated, most valves are oversized, mostly due to overestimating pipe pressure losses that tend to lower the sizing ΔP and by allowing for future increases in throughput. Typically, you wind up with using only 40% of the valve's rated C_v. A linear valve will be about 40% open (using only 40%

of the controller's signal span). An equal percentage valve, on the other hand, will be at about 75% travel. Now, which one do you think will make the better optical impression on your supervisor? Never mind also that the valve's dead band is now spread over 75% of signal span instead of only over 40%. You guessed it: the valve with the equal percentage plug!

Let's consider another argument. Assume the liquid flow through a heat exchanger may vary 3:1 and the corresponding ΔP across the valve, typically to the square of the flow rate, i.e., nine times the ΔP at 1/3 the flow. Our rangeability, or turn-down, i.e., ratio of maximum to minimum C_v is now: $[1/(1)^{1/2}]/[0.333/(9)^{1/2}] = 1/0.111$ or 9:1. Dividing the 40% of rated C_v at maximum flow (see above) by nine now yields 4.4% of rated C_v at minimum flow.

For a valve with a linear characteristic, this occurs at about 3% travel or in the "danger" zone (below 5% travel) where the plug may bump against the seat following a slight controller instability. The equal percentage valve, on the other hand, has about 14% travel at the minimum flow! These are arguments you should keep well in mind.

With the availability of self-tuning controllers and computer chips that can vary the shape of the output signal curve to your valve, the question of what to choose as an inherent valve flow characteristic becomes more irrelevant. Keep in mind that most rotary control valves have "natural" inherent characteristics that cannot be defined in either of the classic definitions, yet they all seem to work well.

This all proves how forgiving most control loops are. Remember, any flow characteristic is linear when viewed under a magnifying glass. For continual processes where the flow rate does not vary by more than 20%, the gain change is less than 2:1, regardless of the type of inherent characteristic, so you might as well choose equal percentage because of the advantages outlined above.

A final word regarding the use of characterized cams in positioners attached to rotary control valves. These are of limited use. First, if the valve has a nearly linear inherent characteristic, then 4% C_v is still only 3% travel, regardless of the 14% signal input required by the equal percentage cam. Second, the modified relationship of travel to signal is correct only under static conditions, i.e., it only works with large process time constants such as with temperature control. Finally, characterized cams are inaccurate and difficult to adjust.

If the process is fast, you have a control valve that has a variable time constant with travel, which can raise cane with the stability of the loop. Why? Because if you have an equal percentage cam to convert (for example, a linear inherent valve characteristic), it requires the valve actuator to move about 21% of its rated travel for a signal change from 90% to 80% but only 1.6% of travel to go from 20% to 10% of signal. Expressed differently, it takes 13 times longer for the valve to move at the top of the signal change than near the bottom of the signal span! This can drastically interfere with the time constant of your process loop. Therefore, avoid cams for pressure or flow control (fast loops).

HOW TO CALCULATE THE "INSTALLED" FLOW CHARACTERISTIC

Before you sign a purchase requisition for a given control valve, you may want to know if your controller remains stable in a control loop where the flow rate can vary over a wide range allowing only the maximum control valve gain change of ±50% mentioned earlier as a criterion. However, this cannot be done without calculating or plotting the installed flow characteristic of the valve and piping system combination, which, when completed, should look similar to the curve in Figure 8-2.

Let's assume we have a liquid flow system that consists of a storage tank, a centrifugal pump, the control valve, more than 20 feet of 6 inch pipe, four elbows, and a process vessel. In order to do the job, we need the following information.

- The maximum and minimum flow rate, given as 1250 and 250 gpm respectively.
- The pump characteristic (converted from feet to psi), given from the manufacturer as:

gpm:	1250	1050	850	650	450	250
psi head:	110	121	131	140	146	150

- The friction loss in the pipe and elbows, calculated as 28 psi at a flow rate of 1250 gpm. Assuming there is no problem with viscosity or flashing, we can assume that the friction loss is proportional to the square of the flow rate, i.e., ΔP pipe = 28 psi/(gpm/1250 gpm)^2; therefore:

gpm:	1250	1050	850	650	450	250
ΔP pipe:	28	19.8	12.9	7.6	3.6	1.1

- The static head change (elevation difference) from pump intake to process vessel (converted from feet to psi), given as 23 psi (remains constant).
- The backpressure in the process vessel, given as 52 psig (remains constant).

First, we find the available pressure drop across the control valve at maximum flow (1250 gpm):

$$\Delta P \text{ valve} = \text{pump head} - \Delta P \text{ pipe} - \text{static head} - \text{backpressure} =$$
$$110 - 28 - 23 - 52 = 7 \text{ psi}$$

From the process data, we know the specific gravity = 1.1. We now can calculate the maximum required valve C_v using Equation 5-1:

$$C_v = 1250/\sqrt{(7/1.1)} = 496$$

Since no tight shutoff is required, and based on rangeability and cost considerations, we might consider a 6 inch, angle-seating, butterfly valve with a rated C_v of 825. From the manufacturer's catalog, we get the inherent flow characteristic, which is:

% Travel:	10	20	30	40	50	60	70	80	90	100
C_v:	32	80	130	190	265	345	440	535	680	825

This tells us that the valve will be about 75% open at maximum flow – almost perfect. Now let's find the percentage of valve travel for our other flow rates, i.e., the percentage of controller signal, which (since we did not specify a positioner cam) is about identical to the valve travel. Going to 1050 gpm, we again calculate the valve pressure drop as above. This is 121 psi using (pump head) – 19.8 psi (ΔP pipe) – 23 psi – 52 psi = 26.2 psi, and the C_v = 1050/(26.2/1.1)1/2 = 215. Extrapolating from the C_v table of the valve, we find the travel at 43.3%.

Repeating the same process for all other flow rates, we find the following:

Flow:	1250	1050	850	650	450	250
% Signal = % Travel	75	43	31	22	15	9
Valve C_v:	496	215	136	90	57	29

The flow vs. % signal is now our installed flow characteristic as shown in Figure 8-3, Curve A, which turns out to be highly non-linear with a gain variation of about 6.7:1 between low flow and high flow! This obviously will result in an unstable controller situation. Unfortunately, we are stuck with the inherent characteristic of the butterfly valve. But even an equal percentage globe valve would not have been much better with the pressure drop across the valve varying over a ratio of 74 psi/7 psi = 10.5:1. Incidentally, if the line size would have been 8 inch, the required reducers would have distorted the installed characteristic even worse. Another argument for line-size butterfly valves!

So, what can we do? A quick check in a popular control valve manufacturer's catalog shows the maximum C_v of a 6 inch cage valve with equal percentage trim is 357, which is too low. An 8 inch size valve is available with a rated C_v of 570. This would fit the bill, but the cost is high and it would be awkward to install a valve larger than the pipe size. Besides, the characteristic would not have made that much of an improvement.[2]

A cam modified positioner characteristic is of no use for other than a slow temperature control system (see Chapter 9), so we have to rely on the modern miracle of electronics. Let's assume we modify the output signal (S_i) from a computer instead of sending it directly (via a digital-to-analog converter) to the valve in the standard 4-20 mA range. If S_i is the signal generated by the internal control-

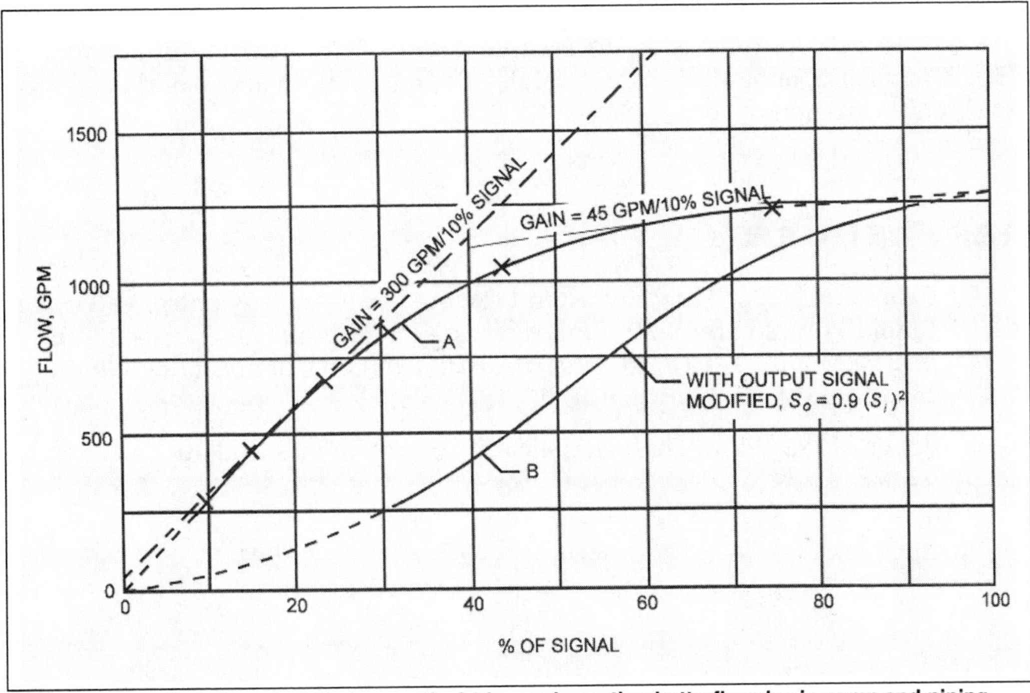

Figure 8-3. Installed characteristic (curve A) of 6 in. angle seating butterfly valve in pump and piping system having a pressure drop varying between 7 and 74 psi. Curve 8 represents the same system characteristic when internal controller logic signal of computer *(S_I)* is modified before being converted to analog 4-20 mA output signal S_o. *Note: digital positioners need no conversion to analog.*

ler logic and S_0 is the final output signal to the valve, we can program the following modification in this case:

$$S_o = 0.9\,(S_i)^2$$

In this particular case, I use 0.9 to bring the valve travel closer to the maximum signal and use the square function to linearize the shape of the installed flow characteristic. This yields the following relationship:

S_i%:	10	20	30	40	50	60	70	80	90	100
S_o%:	1	4	8	14	23	32	44	58	73	90

We can now re-plot the resultant characteristic, for example, by taking the flow at 32% travel (S_0) from Curve A (880gpm) moving it to the corresponding 60% computer signal (SJ point. After transferring all points, we now have the relationship between flow and controller logic signal (S_i) in the form of Curve B. While we still have some gradual change in slope, the resultant gain* variations

* The control valve gain is the percent of flow change divided by a percentage of signal change, Thus, a gain of 1.35, in this case, will yield 13.5% of flow (125 gpm) increase per each 10% of signal increase.

from 1.92 to 0.72 (240gpm/10% to 90gpm/10%) are all within the stated limitation of ±50%. If the controller is set to a valve gain of 1.35, then the upper limit = 1.35+50% = 2.025, and a lower limit = 1.35-50% = 0.68, both above and below the actual gain variations.

REFERENCES

1. Baumann, H.D., "How to Assign Pressure Drop Across Control Valves for Liquid Pumping Systems," *Proceedings of the 29th Annual Symposium on Instrumentation For The Process Industries,* Dept. of Chemical Engineering, Texas A&M University, Jan.16, 17, 18, (1974).
2. Boger, H.W., "Flow Characteristic for Control Valve Installations," *ISA Journal,* pp. 50-54, November (1966).

9

VALVE POSITIONERS

WHEN TO USE VALVE POSITIONERS

While Positioners complicate over all loop stability, most control valves can not operate without them. Broadly speaking, use of a valve positioner, which really is a "valve stem position controller" adds another control loop within a control loop. Here the set point is the controller signal to the positioner; the feedback comes from the lever attached to the actuator stem sensing the latter's position; the controller is the flapper-nozzle, or pilot valve, sensing the difference between the force created by the controller signal and that of a feedback spring attached to the lever; and the final control element is the relay that sends air pressure to the valve actuator. While the latter description refers to pneumatic positioners, the basic functions are identical for digital positioner, except here the signal created input force comes from a voice coil, a torque motor, or a solenoid.

Like any other loop, the positioner loop has its own dead time, time constant, and gain, which quite often interfere with the original process control loop dynamics [1]. This can make the task of the process controller more difficult when fast loops, i.e., for pressure or flow control are encountered.

So much for outlining the problem, now to the benefits. First, there are instances in which the use of a valve positioner is unavoidable:

- When a cylinder actuator is used that requires air pressure in excess of 15 psi. Here, the positioner is used both as a pressure multiplier and as a stem position controller.
- When a valve is split ranged, i.e., 3-9 or 9-15 psi (4-12 mA or 12-20 mA) signal to two valves from a common controller. The preferred way is to use two separate electronic signals from a common controller (use two I/P transducers for pneumatic actuators).
- When valves (usually one of two) are operated in a reverse acting mode, i.e., when a valve, which for safety reasons fails

closed (air-to-open actuator), shall be operated in the signal-to-close direction.

- When the expected dead band of the actuator/valve combination exceeds 5% of signal (0.6 psi or 0.8 mA or equivalent bi-directional electronic signal) span because of excessive stem or trim friction.
- When a booster has to be used for high travel speed and the valve may have excessive dead band.
- When you want to have diagnostic capabilities on line (smart positioners only).

I am not aware of any other reasons why a valve positioner should be used except perhaps to increase the speed of response in case a valve is located too far away from a control room installed I/P transducer. But even there, a valve-mounted transducer or booster relay will perform better with an ordinary valve not meeting any of the criteria above. Think also of how much money will be saved not only in hardware but in piping, air sets, and maintenance as well!

However, since a positioner is required in most cases, try to specify a modern version that offers the ability to vary its (and that of the connected valve's) operating characteristic. Formerly, valve positioners were like process controllers without proportional band and reset adjustments. One may get lucky and everything would be stable; then again, the valve (or the complete control loop) might "hunt" in an unstable mode. It used to be educational to walk through a typical refinery or chemical plant and count all the control valves that are in operation *with the valve positioner disconnected or in the bypass mode!*

To make life easier for today's process operators, many new positioners offer speed and gain adjustments or a combination of the two. This is like adding the adjusting knobs back onto the controller housing. This enables you to change the frequency response characteristic (or time constant) of the final control element in optimum relation to that of the process. You probably know that a loop cannot be stable if the process time constant is equal or close to that of the controller/valve combination. In order to have a stable loop, the ratio between the valve and the system time constants should be at least 3:1 (or 0.3:1) – better yet, 5:1 (or 0.2:1). Ideally, the valve/positioner loop should be faster than the process loop, which is sometimes hard to accomplish.

With a conventional (and non-digital) valve/positioner combination, you had no choice but to either reduce the controller settings (high proportional band – long reset action) or crimp the output tubing from the positioner to the actuator in order to slow the valve down, particularly when the positioner was unstable due to excessive valve stem friction. Now, hopefully avoiding the pitfalls of having somebody ignorantly playing with the adjustments in the field, you are able to change the speed of the valve travel (typically by 5:1) and/or the open-loop gain* setting (sensitivity) of the positioner (typically by 10: 1) even while the valve is in operation. (A word of caution: you may have to re-align the zero adjustment somewhat after changing the gain setting.)

A typical example of how the time constant of the valve can be varied is shown in Figure 9-1. With maximum air flow and a lower gain setting, the "break frequency" is 0.75 cps. At the other extreme, an air flow of only 0.6 m^3/h and a higher gain of 67 (equivalent to 1.5% proportional band) produces a break frequency of 0.32 cps—almost a 220% change in the time constant. The time constant of a control valve is defined as the time in seconds that it takes to travel from zero to 66.7% of rated travel when the full signal is applied. Here is a tabulation of typical time constants for control valves with positioners:

Time Constant(s)

Actuator Size	Rated Travel	Air In	Air Exhaust
32in^2	1/2 in.	1.1	2.5
32in^2	3/4 in.	1.8	2.6
54in^2	3/4 in.	5	7
108in^2	2-1/4 in.	10	12

Even if you don't have the faintest notion about your process dynamics, keep this in mind:

- If you have a slow process such as temperature control, set the gain at or near maximum. This way the valve will respond to the smaller incremental signal change (due to reset action of the controller). However, set the speed (air output) near minimum, thus avoiding an unstable actuator.
- If you have a fast process such as pressure or flow control, set the gain near minimum and set the actuator travel speed at its maximum value, thus creating a very responsive but stable valve.

WHAT TO LOOK FOR WHEN SPECIFYING POSITIONERS OR TRANSDUCERS

The majority of all control loops use a 4-20 mA analog or digital signals, so you are forced to transduce the electronic signal to a pneumatic one, or use a digital

---

* The definition of the "open-loop gain" of a positioner is the change in output signal over the change in input signal (when the feedback is disconnected). Thus, a gain of 200:1 causes a 20 psi output signal change per 0.1 psi input signal change. Gain is the reciprocal of percent band; hence, a gain of 200:1 is equivalent to 0.5% proportional band. Some positioners have gains only as low as 30:1; however, some exceed 200:1. Some digital positioners which employ push-pull solenoid valves for I/P conversation may need a lower gain setting in order to avoid "chattering" of these internal devices.

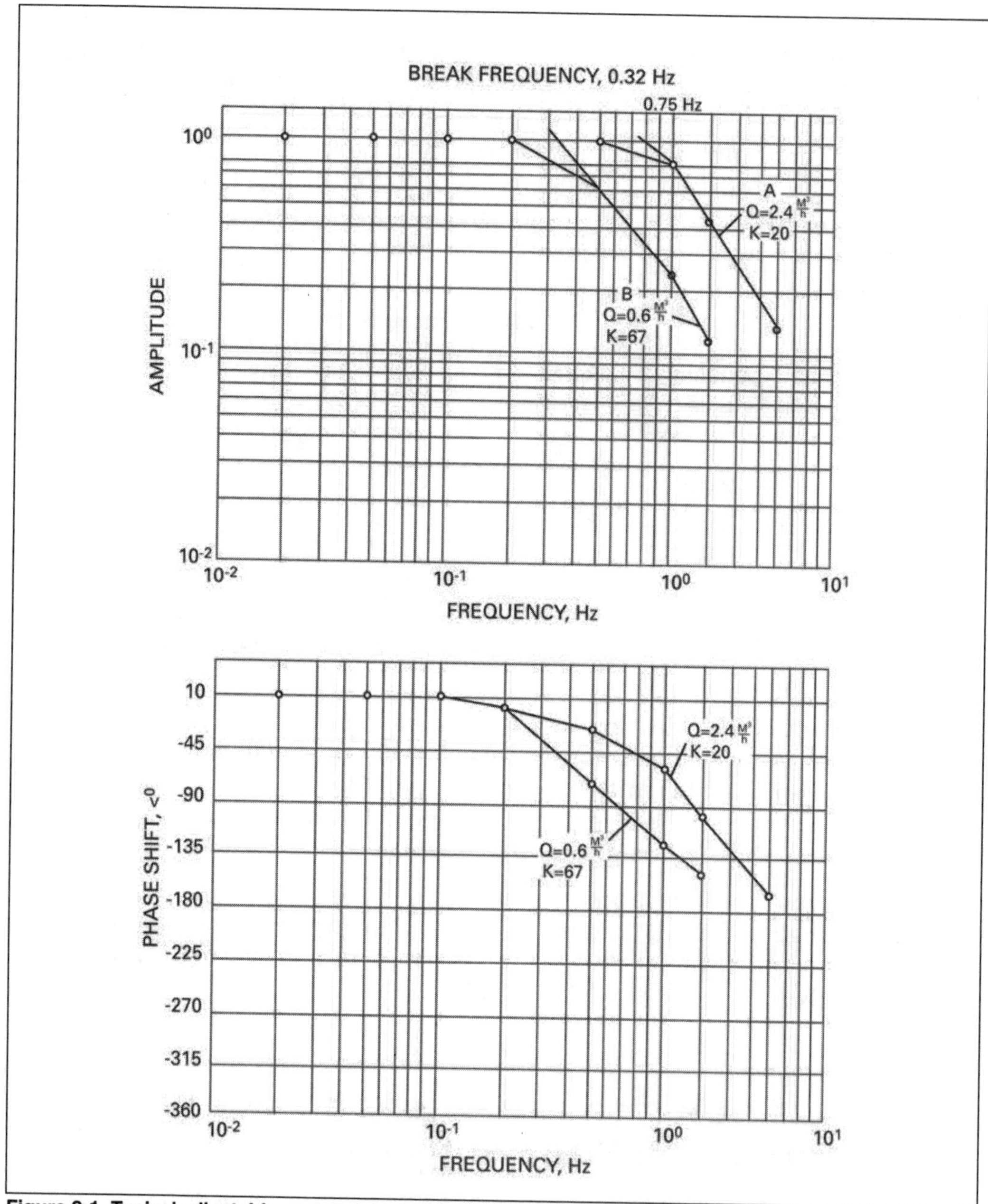

Figure 9-1. Typical adjustable range of dynamic performance of actuator/positioner combination with valve positioners which have separate gain and speed adjustments. (Courtesy of SAMSON AG).

positioner. If you have to use an analog signal (4-20 mA), then an I/P positioner is your choice; alternative, an I/P transducer can be specified who's output then goes to a pneumatic positioner. Note: While pneumatic positioners may be on the way out, they still have cost advantages and are, by design, explosion proof.

First, consider whether or not the valve is installed in an area where flammable liquids are handled. You may need an explosion proof (or intrinsically safe) version. Second, check the vibration level at the valve location. For example, if the valve is close to a centrifugal pump, pipe vibrations can exist in the 60 Hz frequency range, or multiple modes thereof. Any good I/P positioner or transducer should maintain steady output under vibration levels of at least 3g's at 30 cps and 1g at 100 cps (g = acceleration of gravity, 31.1 ft/s^2). Third, consider the quality of air supply. Do you use dehydrated and oil-free instrument air, or air taken directly off the factory compressor? In the last case, avoid all positioners that utilize built-in amplifying relays. The small passages may be subject to plugging. Fourth, consider the effects of weather, if an outdoor installation, or high ambient temperature, if the valve is located near a furnace, for example. Finally, where good frequency response is a must (flow control, turbine or compressor bypass, pressure control, etc.), do not use a I/P transducer that is rack-mounted on a remote instrument panel. Instead mount the I/P transducer on the valve actuator to avoid signal delays.* Make sure the air output capacity of the transducer or positioner is sufficient to give adequate travel speed. Note: Some rack-mounted I/P transducers have air capacities of 1-4 scfm, which limits their use to temperature applications and to the operation of smaller valves only (if no positioner is used).

As a rule of thumb, allow at least one scfh at 20 psi supply for each in.^{2} of diaphragm area (3-15 psi signal). For example, a valve with a 100 in.^{2} diaphragm actuator should need a positioner having at least a 100 scfh air output capacity. For very large actuators, you may need a booster relay between the positioner and actuator. Here, a 1:1 pressure ratio volume booster with a 1 psi differential and adjustable bypass needle valve is recommended. The 1 psi differential ensures that the booster is only activated when the signal output from, say, the positioner, jumps by more than 1 psi. This is usually the case if the valve should open fast in the event of an emergency (typical compressor surge control application). The adjustable needle valve can be tuned to avoid instability of the booster relay.

Other desirable features are: all exterior linkages to be type 18-8 stainless steel; all aluminum parts to be anodized (interior and exterior) *and* epoxy coated; positioners to have at least a built-in gain adjustment; I/P transducers to have the capability for 1-17 psi air output range (see Chapter 10 on off-balance plug forces); and, if of "closed-loop" design (with built-in pressure transducer feedback), should have a gain or dead band adjustment to avoid instability when mounted close to a small diaphragm actuator.

SMART POSITIONERS – SMART VALVES

As is the case with many marketing catchwords, the term "smart positioner" is not yet well defined. It generally is the equivalent of a remote programmable transmitter using the equivalent HART™ protocol or a bi-directional digital sig-

---

* This is especially important for "fast loops" such as flow or pressure control applications, since time lags due to long transmission lines between transducers and valve actuators may vary from 0.2 to 2 seconds per 10% signal change.[2]

nal from a fieldbus device. Nevertheless, most are still working internally with a 4-20 mA analog signal.

On the input side, there are digital electronics that allow for programmable input, such as the inherent characteristic between signal and valve travel (note the resultant nonuniform valve positioner time constant; refer to previous subchapter), the adjustment and calibration of valve travel, split-ranging, and dead band. Nevertheless, the final output signals to the air conversion mechanisms are usually in analog form. Most of these devices do not need external power and are, therefore, intrinsically safe (explosion proof). Fieldbus positioners can perform the functions of a PID controller (act as a decentralized controller with inputs from digital transmitters or computers). Failure of the electronic positioner circuit affects only the local loop.

While these positioners cost about one-third more than their analog counterparts, they are capable of monitoring the control valve. For example, they can tell the computer the current travel position, and by measuring the relationship between the positioner signal and air output pressure to the actuator, they can identify the onset of too much dead band (if, for example, a maintenance person tightens the stem packing too much). On the other hand, a remote calibration of the positioner's zero and span setting after installation, while touted as a big feature, will hardly ever happen.

Of more interest is perhaps the hardware in these positioners. In one particular model, a microcontroller commands one piezoelectric valve to open admitting air to the actuator upon an increase in signal and can command a second piezoelectric valve to open, which exhausts the air from the actuator upon a decrease in signal. If there is no signal change, then both piezoelectric valves stay closed; hence, the claim of low air consumption. These devices require a minimum dead band in order to prevent "hunting" of the opposing valves. However, the majority of smart or digital positioners employ build-in I/P transducers in order to generate the pneumatic output signal. All of these positioners can operate from the current 4-20 mA signal. No auxiliary power is required.

Digital positioner electronics can include PID (proportional, integral, and derivative) control circuitry to do the primary control functions (the computer will do the supervisory override and the changing of set points). Future versions may offer more advanced control functions such as "fuzzy Logic". Be aware of the local environmental situation at the valve's location before using the positioner as a local controller. For example, the valve may be situated in a boiler room with a hot and humid atmosphere, or, the valve may handle a high pressure drop with the associated high vibration level, all potentially affecting the reliability of the control loop. Some vendors offer remotely mounted positioners using electronic stem position measuring devices that tells the positioner the valve's travel position.

Using the valve as a decentralized control system is already done in some "smart valve" designs (shown schematically in Figure 9-2), which also include a built-in flowmeter consisting of piezoelectric pressure-measuring elements upstream and downstream of the valve orifice plus a temperature sensor. Knowing the orifice area (Cv) at a given travel position, the differential pressure across it, and the density (function of pressure and temperature), the flow rate across the valve can be calculated and controlled (for a similar schematic, see Figure 9-3).

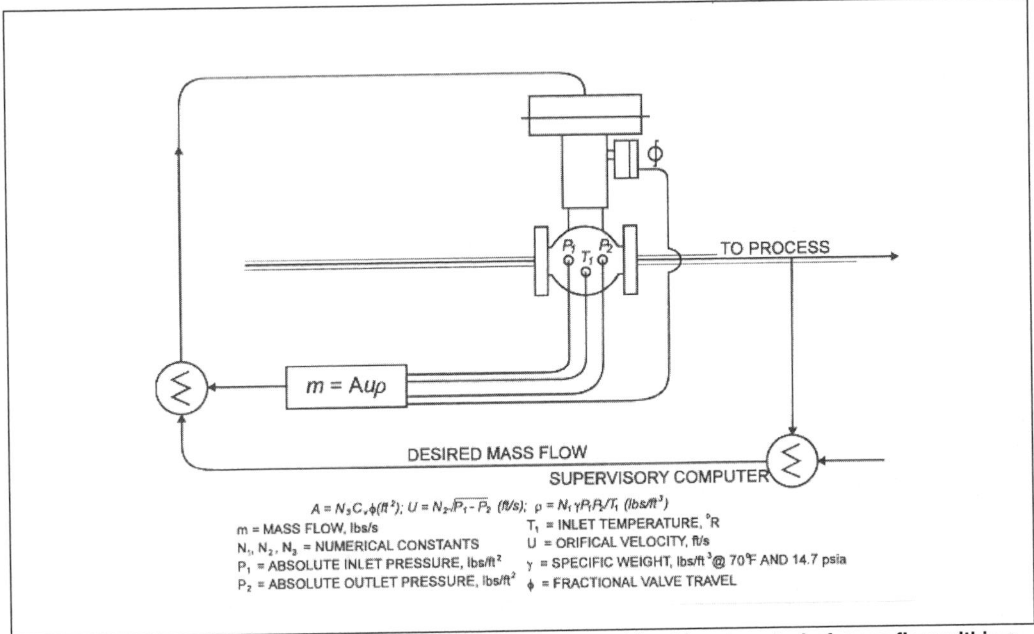

Figure 9-2. Functional diagram of a "smart control valve" capable of local control of mass flow within a cascade control system (Courtesy of FLOWSERVE INC.)

Smart valves have a number of advantages, first they offer a higher range of flow rates (up to 100:1) compared to about 25:1 using flowmeters; this is due to the fact that the movable valve plug acts as a variable orifice. Secondly, the use of smart valves will eliminate the errors produced by converting from analog-to-digital signals. Finally they offer a faster response to loop up-sets.

While much more expensive than their conventional counterparts, such valves can be an important element in more complex control schemes; for example, when you want to control the temperature in a fractionator (a temperature control system with long time constants and time lags), in which the heating medium is steam. This needs a pressure or flow control system which interacts with the temperature control in a typical cascade system. Here the smart valve is the ideal choice for controlling the supply of steam to the heat exchanger. While the temperature controller experiences time lags (typical for temperature changes), the smart valve controlling the steam can react nearly instantly to any up-sets in say the local steam pressure.

Remember, even the smartest valve is no substitute for a flow meter. The variable geometry of the trim, wall attachment phenomenon (Conada effect), incipient cavitation, choked flow, and last but not least, the noise picked up by the pressure transducer due to valve turbulence, make it much too inaccurate. Nevertheless, this does not matter for control. Here it is not the absolute value of the flow rate that counts, but the *deviation* thereof that has to be controlled.

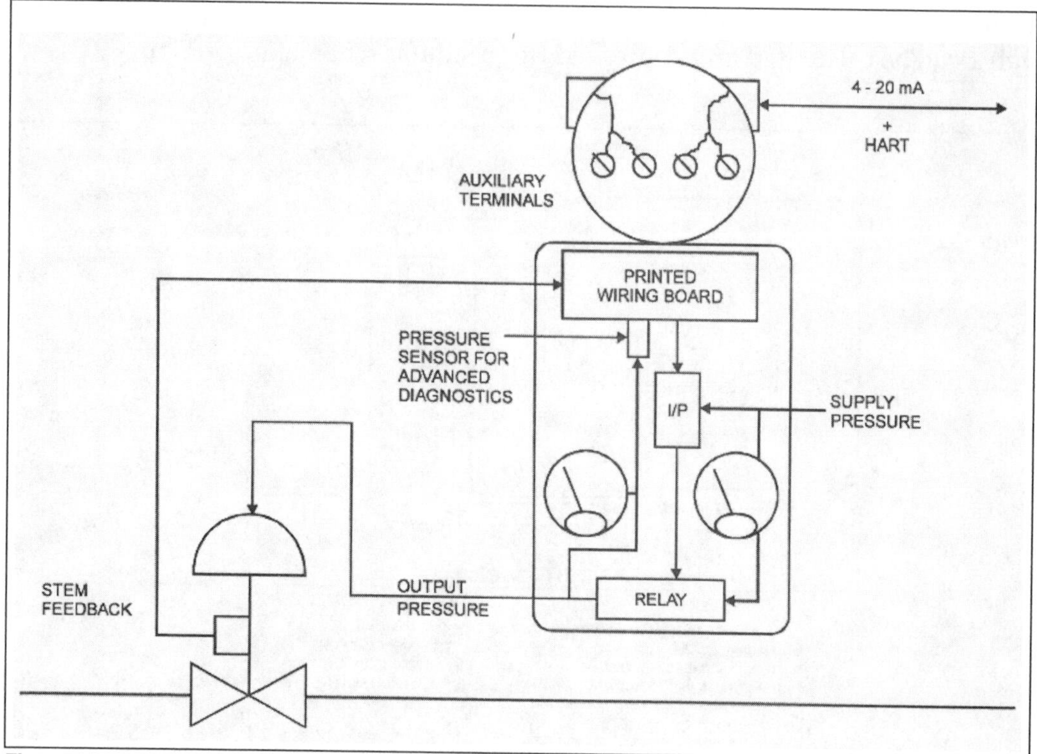

Figure 9-3. Principle of operation of a typical intelligent digital valve positioner.

As part of the progress experienced in the automated control field, smart or digital positioners enhance the function of all currently available control valves and, therefore, offer a higher degree of flexibility. Figure 9-3 shows a schematic view of such a positioner. Communication from a supervisory computer can travel over the present 4-20 mA signal wires or through fieldbus wiring. These same wires feed back information regarding the valve's status to a computer or a display panel. This "feedback" information can be transmitted in digital form via HART™ protocol over the analog 4-20 mA signal. This can be done even while the valve is operating, although special software is required on the receiving end to decode the messages. In fully digital communication protocols such as the Fieldbus Foundation™, both signal and return information will be transmitted in digital form.

The printed circuit board in Figure 9-3 acts as the onboard controller and signal processor. This is typically an encapsulated and field-replaceable element. The circuitry compares this signal level to the signal coming from the stem position sensor, and if not identical, puts out a corrective signal to the I/P transducer element whose output is amplified in a separate relay before being fed to the pneumatic valve actuator as "output pressure" in order to correct the stem position error.

A separate pressure sensor is a key element for the integral diagnostic circuit. For example, by measuring the required output pressure to move the valve stem to, say, 50% travel, and then comparing it to the same pressure (at 50% travel) measured when the valve was brand new, one can detect wear or blockage, or perhaps a too-tight packing box. Such data is fed back, as previously explained, in order to alert the operating personnel and to perhaps initiate maintenance work on the valve. Such preventive maintenance testing can be done either after the process has been shut down, or in the field under actual operating conditions (*in situ*). The latter has the advantage of experiencing actual working conditions such as local temperature and pressure environment. However, such testing can only be done by moving the valve stem in small increments in order to avoid large upsets in the controlled variable. Understandably, this is not widely practiced.

Here is the summary of digital devices compared to analog field devices taken from Reference 3:

1. Increased accuracy, 0.1% to 1% compared to analog 0.3% to 2%.
2. Improved stability, about 0.1% compared to 0.1% to 0.75% analog.
3. Wider range ability (without sacrificing accuracy) up to 50:1 compared to 10:1.
4. Capable of performing multi-functions (see above).
5. Diagnostic and self-testing.
6. Ability for miniaturization.
7. Ability to calibrate and adjust without physical access to the instrument.
8. Utilize physical effects that could not be measured or assessed by analog means.
9. Finally, digital processing can be done with very little energy consumption allowing intrinsic safety under the "triade" policy encompassing U.S/C.S.A. as well as CENELEC (European) and CESI (Japanese) codes.

ACCESSORIES AND SOFTWARE

In addition, these positioners require software in order to perform the required diagnostic or remote calibration functions. Here are some capabilities that such a software package should support[4]: Common accessories for digital positoners are build-in micro switches. These typically operate using a low voltage (24 V). Such switches can provide a safety shut-down of the valve, which can be initiated independently from the controller. Also available are digital or analog travel position indicating devices.

- Reading the valve's diagnostic data.
- Supporting the graphical diagnostic trend window display.
- Valve configuration and valve configuration database.
- Monitoring device capabilities.
- Configuration of valve characteristic.
- General process and valve database.

- Security level for user.
- Diagnostic database.

CONNECTING IT ALL TO THE CONTROL SYSTEM

In order to interact properly, the plant's control system needs an "architecture" where the transmitters, positioners, controllers, annunciators, computers, and so on, plus their associated software, are integrated in a type of structure where each of these devices can communicate effectively with each other and perform in an efficient manner.

An efficient field-based architecture can build process management solutions by networking intelligent field devices, scaleable control system platforms, and software. Besides intelligent field devices, e.g., smart positioners, this requires modular performance software. If done properly, such software can redefine management, expand process control, and add solutions to asset management problems. It should:

- reduce installed cost,
- improve performance,
- it should be compatible with existing devices,
- reduce process offsets,
- it should be scaleable,
- reduce maintenance cost,
- improve plant reliability,
- reduce training time,
- should be adaptable to changing processes, and
- reduce cost.

As shown in Figure 9-4, such architecture will effectively change the previous typical 4-20 mA analog signal based "DCS-Centric" system to a new field-based architecture. This will effectively send the operator back from the control room to the plant level. Smart field devices such as smart valve positioners with control functions (PID) can do all of the following on a local level: control of loop, signal processing, data acquisition, analyses, alarm detection, trend collection, and maintenance recording.

ACCESSORIES

According to Bud Keyes[5] this process of transferring functionality out of the central control room and into the field devices is analogous to the computer industry's move from mainframe computing to P.C.-based networks.

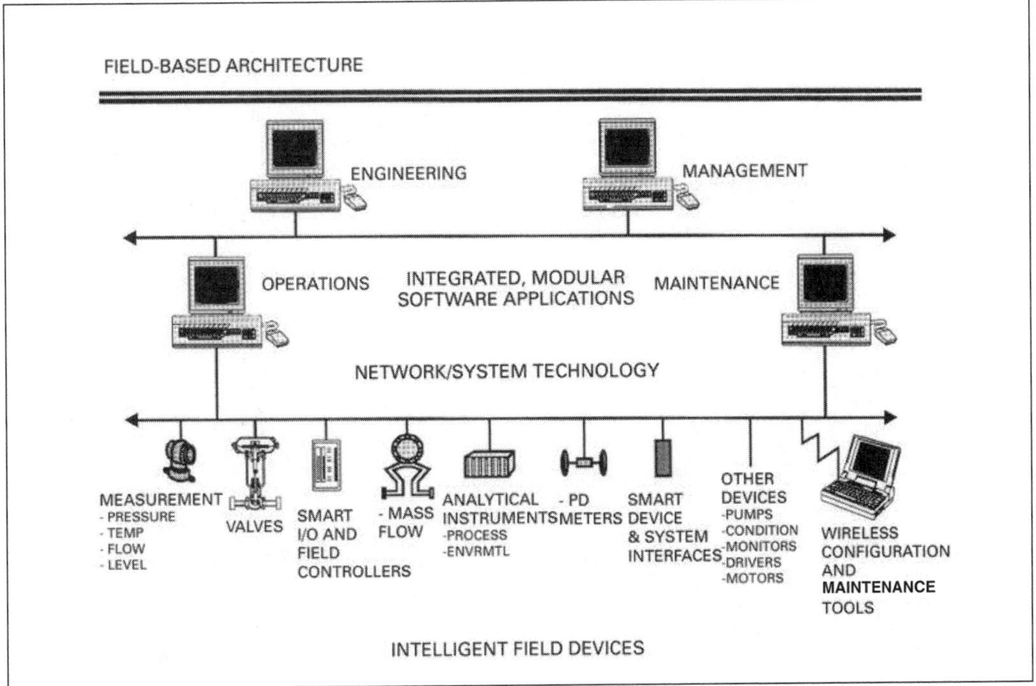

Figure 9-4. Field-based architecture with intelligent field devices. (Courtesy of EMERSON PROCESS CONTROL)

The overall aim is to create a "real time" management control system integrating all aspects of plant management, production control, process control, plant efficiency, and maintenance.

The maintenance program for control valves could include, for example, on-line diagnosis, remote calibration, preventive maintenance plans, and control of spare parts inventory.

REFERENCES

1. Lloyd, S. G., "Guidelines for the Application of Valve Positioners," *Proceedings of the Texas A&M Instrumentation Symposium,* January (1969).
2. St. Clair, D. W., "Controller Tuning and Control Loop Performance," p. 19, Newark, DE: Straight Line Control Company.
3. Schneider, Hans-Josef, "Digitale Feldgerate: Vorteile, Probleme und Anforderungen aus Anwendersicht," *atp - Automatisierungstechnische Praxis,* April (1995), pp. 50-54.
4. Lipták, Béla, G. "Process Control and Optimisation," Fourth Edition, Vol. 2, Taylor & Francis Co. N.Y. 2006, pp. 1197.
5. Keyes, M.A., "Field-based Distributed Control Opens Door to Performance Improvements," *I&CS,* January 1997, pp. 25-35.

10

THE MYSTERY OF LINE PRESSURE-PRODUCED VALVE STEM FORCES, OR SELECTING THE CORRECT ACTUATOR SIZE

The selection of an adequate valve actuator is usually left up to the vendor. Most of the time this works well, especially if the pressure conditions the valve has to handle are low or the valve size is small. However, you may want to know some of the consequences in selecting the wrong (usually too small) actuator size. Remember, the valve vendor who got the job was probably the lowest bidder and, quite often, the bidder who cut the size of actuators in order to reduce costs. What follows will tell you what can happen after installation.

THE VALVE DOES NOT CLOSE PROPERLY

The economically dictated demise of bypass installations around a control valve has led to the requirement that a control valve should have ANSI/FCI Class IV seat tightness (to double as a shutoff valve when the system is down). This works well as long as the initial spring setting, the air pressure, and size of the actuator are adequate to overcome the fluid pressure acting on the valve plug (assuming a single-seated globe valve). To start with, it is important to tell the vendor what the maximum inlet pressure is! This is usually the pressure differential across the plug when the valve is shut off and *not* the lower inlet pressure and substantially higher outlet pressure that the valve sees during normal throttling or that is used for calculating the valve C_v. A responsible valve vendor will always assume the valve has to shut off against the maximum inlet pressure and with a zero gage outlet pressure. Therefore, P_1 has to be stated on the specification form (see Figure 10-1).

CONTROL VALVE DATA SHEET

ISA		
PROJECT	DATA SHEET	OF
UNIT	SPEC	
P.O.	TAG	
ITEM	DWG	
CONTRACT	SERVICE	
MFR. SERIAL*		

1 Fluid Crit Press PC

#			Units	Max Flow	Norm Flow	Min Flow	Shut-Off
2	SERVICE CONDITIONS	Flow Rate					—
3		Inlet Pressure					
4		Outlet Pressure					
5		Inlet Temperature					
6		Spec Wt. / Spec Grav / Mol Wt.					—
7		Viscosity / Spec Heats Ratio					—
8		Vapor Pressure P_v					—
9		Required C_v*					—
10		Travel*	%				0
11		Allowable / Predicted SPL*	dBA				—
12							

#	Section	Item	#	Section	Item
13	LINE	Pipe Line Size In	53	ACTUATOR	Type*
14		and Schedule Out	54		Mfr & Model*
15		Pipe Line Insulation	55		Size* Eff Area
16	VALVE BODY / BONNET	Type*	56		On / Off Modulating
17		Size* ANSI Class	57		Spring Action Open / Close
18		Max Press/Temp	58		Max Allowable Pressure*
19		Mfr & Model*	59		Min Required Pressure*
20		Body / Bonnet Matl*	60		Available Air Supply Pressure:
21		Liner Material / ID*	61		Max Min
22		End In	62		Bench Range* /
23		Connection Out	63		Actuator Orientation
24		Flg Face Finish	64		Handwheel Type
25		End Ext / Matl	65		Air Failure Valve Set At
26		Flow Direction*	66		
27		Type of Bonnet*	67		Input Signal
28		Lub & Iso Valve Lube	68	POSITIONER	Type*
29		Packing Material*	69		Mfr & Model*
30		Packing Type*	70		On Incr Signal Ouptut Incr / Decr*
31			71		Gauges By-Pass
32	TRIM	Type*	72		Cam Characteristic*
33		Size* Rated Travel	73		
34		Characteristic*	74	SWITCHES	Type Quantity
35		Balanced / Unbalanced*	75		Mfr & Model*
36		Rated* C_v F_L X_T	76		Contacts / Rating
37		Plug / Ball / Disk Material*	77		Actuation Points
38		Seat Material*	78		
39		Cage / Guide Material*	79	AIRSET	Mfr & Model*
40		Stem Material*	80		Set Pressure*
41			81		Filter Gauge
42			82		
43	SPECIALS / ACCESSORIES	NEC Class Group Div	83	TESTS	Hydro Pressure*
44			84		ANSI / FCI Leakage Class
45			85		
46			86		
47					
48					
49					
50					
51					
52					

Rev	Date	Revision	Orig	App

* Information supplied by manufacturer unless already specified

 Second Printing ISA FORM S20.50, Rev. 1

Figure 10-1. Control valve data sheet from ISA-20.50-1981.

For example, a heating valve with a 2 in. orifice handling hot oil will see, at the normal flow rate and 75% valve travel, an inlet pressure of 125 psig and a downstream pressure of 110 psig. The stem forces sensed by the actuator under these conditions are $(2^2 \pi/4) \times (125 - 110) = 47$ lbs.* However, if called to close, the inlet pressure will be 140 psig due to the increase in pressure following the pump's characteristic, while the downstream pressure may be zero. In this case, the minimum stem force will be 140 psi times the seating area of the plug (use for seating diameter at least 1/16 in. plus the orifice diameter) plus the packing friction. Hence, stem force (closed) = $(2.063^2 \pi/4) \times 140$ + packing friction.

Now packing friction can vary considerably; for example, tests with TFE-containing packing material showed that starting friction (or break away friction) was 1.5 times that of dynamic (constant movement) friction.[1] However, as a rule of thumb, use 1.5 x stem area x max. P_1. In this case, assume a 3/8 in. stem, P.F. = $1.5(0.375^2 \pi/4) \times 140$ psig = 23 lbs. Total stem force (closed) is, therefore, 468 + 23 = 491 lbs.

As you can see, there is more than a tenfold increase in actuator force requirement. In case of a piston actuator, and assuming the availability of at least 80 psi air, the minimum cylinder area should be 491 lbs/80 lbs/in.^{2} = 6.2 in.^{2} or a 3 in. piston diameter for "air to close" action. You need a minimum spring force of 491 lbs if the valve should fail close upon air failure.

If you want a spring-diaphragm actuator to fail close and you specify a valve positioner, a 6-15 psi span spring inside a 100 in.^{2} actuator would be adequate. Here, the output signal from the positioner would vary as follows:

- At zero travel (valve shut), the force created by the spring is 6 psi x 100 in.^{2} = 600 lbs. However, since the seating stem force here is 491 lbs (trying to push up), it leaves only 600 – 491 lbs = 109 lbs for the actuator to open the valve. Hence, the air signal is 109 lbs/100 in.^{2} = 1.1 psi.
- Remember, at 75% travel (normal operating conditions) our stem forces are reduced to 47 lbs. At this point, our spring exerts a force of [(15 psi – 6 psi) x 0.75 + 6 psi] x 100 in.^{2} = 12.75 psi x 100 in.^{2} = 1275 lbs. Subtracting from this value the 47 lbs acting up (i.e., against the spring) = 1275 – 47 = 1228 lbs net spring force. This translates into a positioner output signal of 1228 lbs/100 in.^{2} = 12.28 psi.
- We thus have an installed positioner output signal span of 1.1 to 12.28 psi between zero and maximum flow at 75% travel. This is in contrast to a controller signal to the positioner of 3-12 psi. This contradicts a common fallacy that the output signal of a positioner has to match the input signal.

As you can see from the above example, if you were to disconnect the positioner, the valve would never close with a constant 3 psi input signal from an I/P transducer (electro-pneumatic converting instrument). This is exactly what has

* Neglecting the area of the valve stem for simplicity.

happened in many plants that utilize computers or single loop digital controllers for process control.[2] These devices lock in the 4 mA signal at zero process conditions.* If you have an I/P transducer that is calibrated to have a 3-15 psi output range for a 4-20 mA input signal (and most are), you will continuously feed 3 psi to the valve and, as a result, could cause valve leakage. Unfortunately, this situation most often leads to a call to the vendor informing them that their valve is bad (lousy quality, etc.).

What can be done about it? Well, in order of cost effectiveness and engineering preference:

- If you can, reduce the zero setting on your I/P transducer to 1 psi at 4 mA, or specify an I/P transducer with a 1-17 psi output per ANSI/FCI Standard 87-2.
- Ask your valve vendor to select an actuator spring with a 8-15 psi "bench range." (This, in the above example, will give you an installed range of between 8 minus 4.9,† when seated, and 15 minus 0.5 psi,‡ at 100% travel, which is 3.1 to 14.5 psi, close enough to be called 3-15 psi range!)
- Specify a valve positioner.

Please note that simply cranking up the initial spring setting from, say, 3 psi to 6 psi won't do. You will wind up with a bench range of 6-18 psi, and your valve will not open enough to handle flow upsets. (Remember, a standard I/P signal cannot exceed 15 psi at 20 mA input!)

THE VALVE IS UNSTABLE

The instability caused by dynamic fluid forces on a plug, disk, or ball should not be confused with positioner or control loop instability (sometimes hard to distinguish). As was shown in the prior example, forces acting on a valve plug or across a butterfly valve (or rotating plug, for that matter) can vary drastically with flow conditions. Many papers have been written on the subject,[3] most of them requiring you to become a math major! Therefore, I will try to keep the matter simple.

For a valve trim to be stable, the change in stem force per inch of stem motion has to be less than the change in stem force per inch of travel exerted by the actuator. The latter is called the actuator stiffness factor, K_n, and is expressed in lbs/in. For a spring-diaphragm actuator this is simply the spring rate of the actuator spring (or all of the springs combined if multi-spring), ignoring the pneumatic damping.

---

* It is possible to adjust your computer to under- or over-range the signal by 5% or 0.8 mA. However, allowing the signal to decrease below 2 mA can cause failure of certain I/P transducers with dire consequences for your process loop!

† 492 lbs. seating force/100 in.^{2}

‡ 47 lbs. stem force/100 in.^{2}

With piston actuators, things get a bit more complex. Here, it is the force generated by compressing the internal actuator air pressure when the piston is moved over a given distance that determines the stiffness. Obviously, this force varies with stem position, initial cylinder volume *(L)*, and absolute air pressure. As a rule, piston actuators are much stiffer[4] than spring-diaphragm actuators and, therefore, are able to operate single-seated globe valves in the flow-to-close direction, a feat normal spring-diaphragm actuators cannot handle.

The general practice is to install single-seated globe valves in the flow-to-open direction, where the problem of dynamic instability seldom arises. However, it can be a problem in force-balanced cage valves and angle valves handling cavitating or flashing fluids where discharging occurs downstream of the valve plug (flow tending to close).

What happens here is that not only does the effective area of plug increase as you approach the seating point, but, in addition, the downstream pressure decreases rapidly, roughly to the square of the flow rate decrease. We all have experienced this phenomenon in the bathtub: with water in the tub, as we slowly let go of the chain, the rubber plug approaches the drain, but, at about 3/16 in. from the seat, the plug suddenly surges down and slams onto the seat. The same thing happens in a valve. The trick is to have the actuator stiff enough to maintain hold of the valve plug.

As a rule of thumb, consider that the off-balance plug area goes from zero to maximum within a distance that is the lesser of either 1/3 of valve travel or 1/4 of the orifice diameter. Further assume that the differential pressure will increase from $P_1 - P_2$ at normal flow to P_1 maximum within this same distance.

Here is an example: A 3 in. angle valve with a 2.5 in. orifice diameter has a 1/2 in. valve stem diameter. Normal P_1 = 165 psig; normal P_2 = 125 psig; maximum P_1 = 175 psig; flow-to-close actuator travel = 1.5 inches. Since the area of the stem is relatively small, we can ignore this and assume the off-balance plug area is the 2.5 in. orifice area, which is 4.9 in.^{2}.

First consider the distance of area change: 1/3 of 1.5 in. travel = 0.5 in.; 1/4 of 2.5 in. orifice diameter = 0.625 in.; therefore, we chose 0.50 in. being the smaller. The rate of change in stem forces is now

$$\text{Off-Balance Area x } P \text{ change/Distance} \qquad (10\text{-}1)$$

where: ΔP change = max. $P_1 - \Delta P$ or 175 – (165 – 125) = 135 psi.

Therefore, from Equation 10-1, 4.9 in.^{2} x 135 psi/0.50 in. = 1323 lbs/in. Assuming the actuator is spring-diaphragm with an area of 100 in.^{2} and a 3-15 psi bench range, this makes the spring rate (15-3 psi) x 100 in.^{2} /1.5 in. travel = 800 lbs/in. This is quite insufficient, and we either have to recommend a 175 in.^{2} actuator or change the bench range of the spring (or springs) to 6-30 psi. This would increase the actuator stiffness to 1600 lbs/in., which will provide a sufficient safety margin. Air compression within the actuator will increase the stiffness and, therefore, provide a safety factor, although such stiffness is much less than in cylinder actuators due to larger air volume.

Rotary control valves have similar stiffness problems. They are more complex since we have to deal with torque instead of force. Consider the combined torque characteristic of a butterfly valve (shown in Figure 10-2, taken from Figure 4 in Reference 5). Here, the greatest change in torque per degree occurs at 60°, where the combined torque changes at a rate of (0.011 in.-lbs/d^3 psi) x P_1 for ±5° travel, i.e., from 0.064 @ 55° to 0.075 @ 65°. Let's assume we have a 10 in. valve with a maximum differential pressure (when valve is closed) of 87 psig. We have a torque change at 60° opening (and assuming the valve has to open above 60° to pass the maximum flow) of 0.011 in.-lbs x 10 in.^{3} x 87 psi = 960 in.-lbs per 10°.

Assume we have a cylinder actuator with a 75 in.^{2} area that travels 4 inches to move the vane from 0° to 90° opening. Neglecting angular cosine effects, we can assume that at a 65° opening the piston travel will be at (4 in./90°) x 65° = 2.88 inches, and at 55° vane travel = (4 in./90°) x 55°, at 2.44 inches.

What we need to know is the volume change in the actuator going from 2.44 inches to 2.88 inches piston travel and assuming 0.3 inches of over travel (0. T.) in the cylinder. Let's calculate the ratio of air volume between both travel positions:

$$\text{Volume Ratio} = \\ \text{Area x (Initial travel + Over travel)/Area x (Final travel + Over travel)}$$

Therefore,

$$\text{Volume Ratio} = 75 \text{ in.}^2 (2.88 \text{ in.} + 0.3 \text{ in.})/75 \text{ in.}^2 (2.44 \text{ in.} + 0.3 \text{ in.}) = \\ 239 \text{ in.}^3/206 \text{ in.}^3 = 1.16$$

Assuming further that the positioner output pressure in the cylinder at 65° travel is 65 psia, we can now calculate the force change due to air compression, if the actuator should suddenly have to move from 65° to 55° because of dynamic torque effects (remember that in most rotary valves torque tends to close).

$$\text{Act. force change} = (\text{Volume ratio} - 1) \text{ x Area x Air pressure } (10 - 2) = \\ (1.16 - 1) \text{ x } 75 \text{ in.}^2 \text{ x } 65 \text{ psia} = 780 \text{ lbs}$$

All we have to do now is to convert this force change to torque. Since we know the travel is 4 inches at 90° rotation, we can assume that the effective length of the lever turning the shaft is equal to actuator travel/2 x tan45°= 4/(2 x l) = 2.00 inches.

Thus, the force change of 780 lbs translates into a torque change of 780 lbs x 2.00 in. = 1560 in.-lbs/10° travel. This is above the dynamic vane torque change of 960 in.-lbs/10°, and we could expect stability of operation since the friction of the piston seal will provide extra resistance to sudden change.

If this had been a spring-diaphragm actuator with the same 4 in. travel, a torque requirement of 960 in.-lbs per 10° vane travel would demand a 960 in.-lbs/2 in. effective lever = 480 lbs force change per 0.44 in. actuator travel and, therefore, would require a 480 lbs/0.44 in. = 1091 lbs/in. spring rate. Neglecting added stiffness due to air compression, this is equivalent to a 180 in.^{2} actuator with a 6-30 psi bench range. Quite a drastic increase in actuator size.

While the above example assumed a conventional vane, so-called "low torque" vanes would reduce the stiffness requirement and would, therefore, be very effective when allowing for the use of spring-diaphragm actuators despite their lower stiffness factor. Note: A 60° opening with the 15 in. angle-seating, low-torque vane occurs at 45° vane travel (see Figure 10-2).

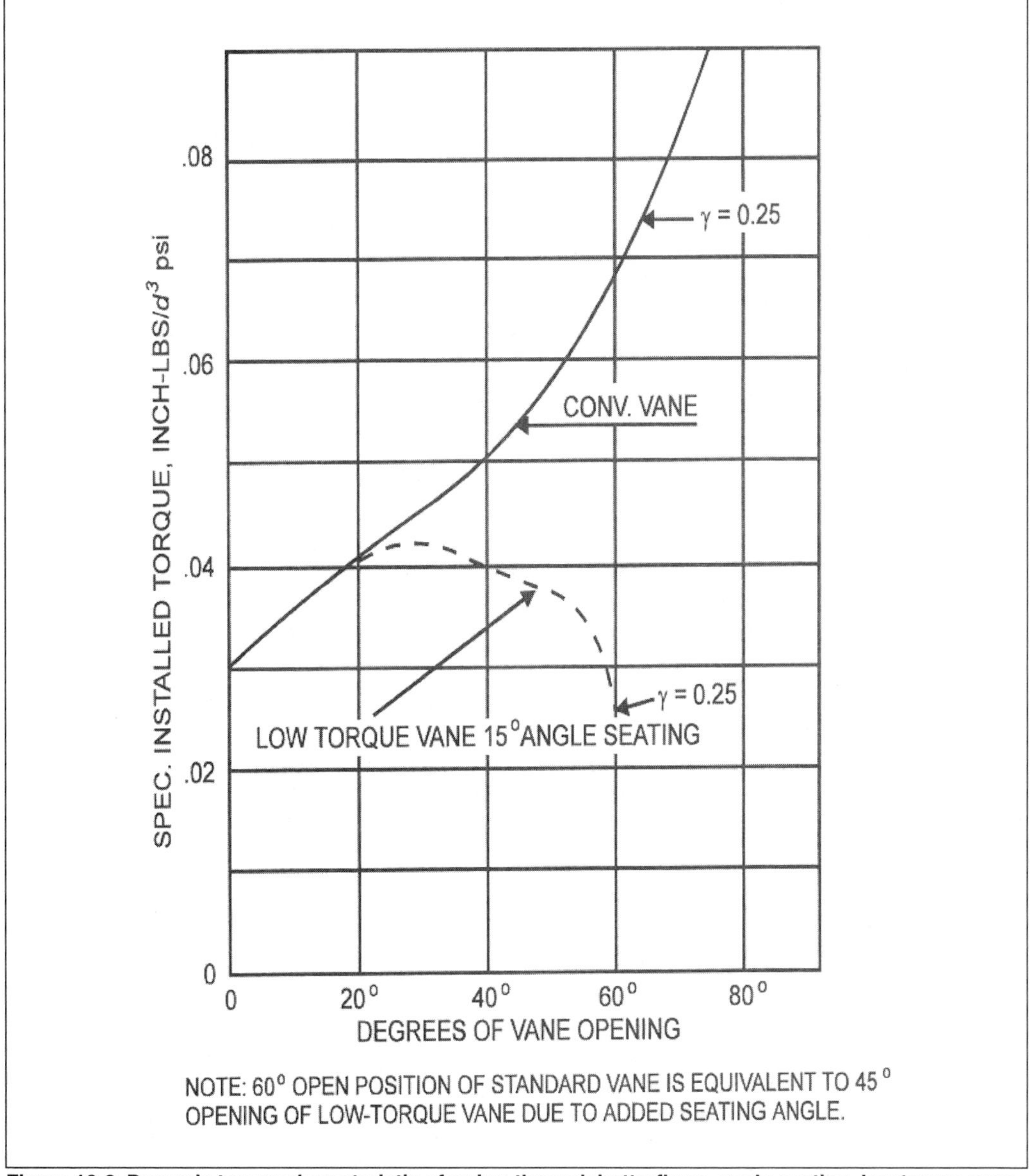

Figure 10-2. Dynamic torque characteristic of swing-through butterfly vs. angle-seating, low-torque vane.

The following is a summary of stiffness equations.

- Spring-Diaphragm Actuator (ignoring lesser stiffness due to air compression*)

$$K_n = (\text{Diaphragm area x Signal span})/\text{Travel} \;\; \text{lbs/in.}$$

- Cylinder Actuator (single-acting)

$$K_n = \{[A_C (h_i + h_o + \Delta h)/ A_C (h_i + h_o)] - 1\} \times S \times A_C /\Delta h \;\; \text{lbs/in.}$$

where:

A_C = effective area of cylinder, in.^{2}

h_i = initial travel position, in.

h_o = overtravel of piston (travel not used), in.

S = signal pressure at given travel position, psia

Δh = change in travel due to disturbance, in.

- *Cylinder Actuator (double-acting).* Results from single-acting cylinder actuator times two.

REFERENCES

1. Jeschke, N., et al., "Untersuchungen der Stopfbuchsreibung an Stellventilen mit pneumatischen Membranantrieb." *RTP,* Volume 3. West Germany, (1968).
2. Baumann, H. D., "The Need for Pneumatic Power Signals for Control Valves," *INTECH,* pp. 34-41, Jan. (1988).
3. Keith, G. A., "Analytical Prediction of Valve Stability," ISA Paper 838-70, Research Triangle Park, NC: ISA, (1970).
4. Schafbuch, P. J., "Fluid Inertia Effects On Unbalanced Valve Stability," ISA Paper 85-0207, Research Triangle Park, NC: ISA, (1983).
5. Baumann, H. D., "A Case for Butterfly Valves in Throttling Applications," *Instrument and Control Systems,* pp. 35-38, May (1979).

* Assumed to be isothermal.

11

HOW TO INSTALL A CONTROL VALVE

Having properly selected a given control valve does not automatically ensure that it will do the right job. Proper installation is almost as important. Unfortunately, this might depend greatly on the experience of the person who designed the piping layout. However, it may not hurt to ask them a few questions regarding the planned control valve installation. If nothing else, it shows him or her that you care. On the other hand, you might just prevent a major mistake you may be blamed for later.

Here, then, are some, but by no means all, of the important considerations regarding control valve installations.

ACCESSIBILITY. Is there sufficient space to remove the actuator when repair is necessary? Can the flange bolt or the tie rods be removed? (This can be a problem, especially with flangeless valves, if the body is installed between pipe reducers.) Can the positioner gages or the valve travel indicator be seen? Can the positioner be adjusted? If a hand wheel is provided, is it in a position so that it can be rotated? Can the pipe leading to the valve be de-pressurized or drained?

CONSIDERATIONS FOR MANUAL OPERATION. In many critical applications you may want to consider a manually operated throttling valve in a bypass around the valve, with appropriate block and drain valves on either side of the control valve that allow for the repair or replacement of the latter without shutting down the process (ascertain if it is possible to run the plant manually).

LOCATION. Is the location of the valve chosen to permit repair? Is the location optimized to help prevent cavitation or, what is worse, permit cavitation to happen in a normally flashing downstream environment because the outlet pipe is too long? If the valve is on the bottom of a steam header, place a steam trap close to the valve **inlet.** Do not place an elbow or pipe tee less than six pipe diameters downstream of a butterfly or ball valve to avoid interference with the valve's flow capacity.

In high-pressure gas or steam pressure-reducing applications, try to have only straight pipes downstream of the valve or use long sweeping elbows to avoid additional pipe noise. Allow for drainage through the valve. This is especially important for sanitary valves and for valves handling dangerous fluids.

Consider the possibility of corrosion caused by leaky flange gaskets or valve stem packings on equipment located *below* the valve. Is the valve located near a source of extreme heat or where the accessories might freeze up in winter? You may have to consider the ambient temperature rating of the actuator or positioner, or heat trace the valve or air lines. Does the valve have to be fire-safe or must the electrical accessories be explosion proof due to being located in an area considered hazardous?

The minimum you should have are certified dimensional drawings from your vendor to make sure the valve fits into the piping. When you purchase a "flangeless" valve make sure the vendor supplies the connecting tie rods and nuts.

Finally: keep good records of what you purchased in order to be able to replace or repair the valve. The original name plate maybe torn off or painted over after a few years.

OTHER THINGS THAT CAN GO WRONG

- The installer forgets to pre-torque the tie rods that hold a flangeless control valve, especially when it will control a cold fluid.
- The installer paid no attention to the flow arrow on the valve body.
- No filter was placed ahead of a valve that has a close clearance valve plug ($C_v < 0.1$).
- The valve body of a butterfly or ball valve was misaligned with the piping flanges. This can reduce the maximum valve C_v by up to 20%.
- No shutoff valve was provided in the pneumatic supply line to the positioner. Neither was a pressure regulator or air filter.
- An elastomer-lined butterfly valve is installed between slip-on type line flanges, causing the liner to fail because it is not properly retained due to the narrow width of the gasket surface if the liner.
- A PTFE-lined, corrosion-resistant valve is installed between plastic-lined, raised-face line flanges, causing the PTFE valve surface to "cold flow" and cause leakage between the body and pipe flanges.
- A bleed or drain valve was not installed in the pipe to relieve the pressure in the line when the control was shut down.

- The valve installer insulated the extension bonnet on a cryogenic valve, causing the packing box to freeze up.
- No protective metal shroud was placed around the tie rods and wafer-style valve body in an area where there is a potential fire hazard.
- The bellows seal (or stem sealing diaphragm) was not removed from a valve during hydrostatic testing of the valve and piping system causing the bellows (or diaphragm) to rupture.
- The positioner of a valve located outside and subject to freezing temperatures was connected to an ordinary shop air line, instead of to dehydrated instrument air supply.

Of course, your primary concern should always be safety. The installation certainly should meet all safety code requirements and should be investigated to eliminate all hazards, including the possibility of causing pollution. Then, consider saving your company a bundle of money by ensuring that this valve will last for at least 20 years *without* being the cause for a very expensive shutdown of the plant or process.

REFERENCES

1. API RP550, 2nd Edit., *Manual on Installation of Refinery Instruments and Control Systems,* New York, NY: American Petroleum Institute.
2. *Valves, Piping and Pipe Line Handbook,* 2nd Edit., Surrey, U.K.: The Trade and Technical Press Limited, (1986).
3. *Control Valves,* Guy Borden Jr. Editor, 1998., ISA, Research Triangle Park, NC. P. 543.

12

HOW GOOD IS THE VALVE THAT I PURCHASED?

This is a legitimate question for buyers and users of control valves. Aside from being able to meet the obvious requirements such as the flange and pressure vessel code requirements and having the right flow capacity and characteristic, there is the important question: How will it perform in my control loop?

Major users have developed internal standards that tried to address this question. Most of these standards stated certain minimum performance criteria such as linearity, threshold sensitivity (usually 0.2% of span with positioner), dead band (maximum 0.5% of span with positioner and 5% without), plus hysteresis.

However, even if you meet all these criteria, you still cannot expect valves from two different manufacturers to act the same when installed in a given control loop. The reason is the interaction of installed flow characteristic, dead time (caused by air volume in the actuator or stem friction), and valve time constant.

To address the question of "loop behavior" more realistically, some manufacturers of control valves have come up with the idea of testing a valve in an actual flow loop[1], typically controlling the rate of flow (using pumped water). Such a control loop is very fast and poses high demands on a valve and its positioning system. The performance criteria is the amount of deviation of the controlled variable from the controller set-point under various flow (load) conditions and varying the loop's time constant and controller gain (proportional band) settings.

The normalized, averaged deviation of the actual versus set-point error, sometimes called "variability," is then plotted against the closed loop time constant at a given load or valve travel setting (see Figure 12-1). Such a test involving several different valves may be called "performance differentiation" and can be supplemented by a static test (open loop control, controller on manual). Here the travel of the valve and/or the change in flow is observed and recorded following a step change in the controller signal (minimum change = 0.5%), see Figure 12-2.

As you can see from Figure 12-1, different valve types, or valves from different manufacturers, behave quite dissimilar even though each of them meets the customer's requested valve requisition criteria. Looking at the graph, you notice that all valve types perform well when the loop time constant is very long, such as in a

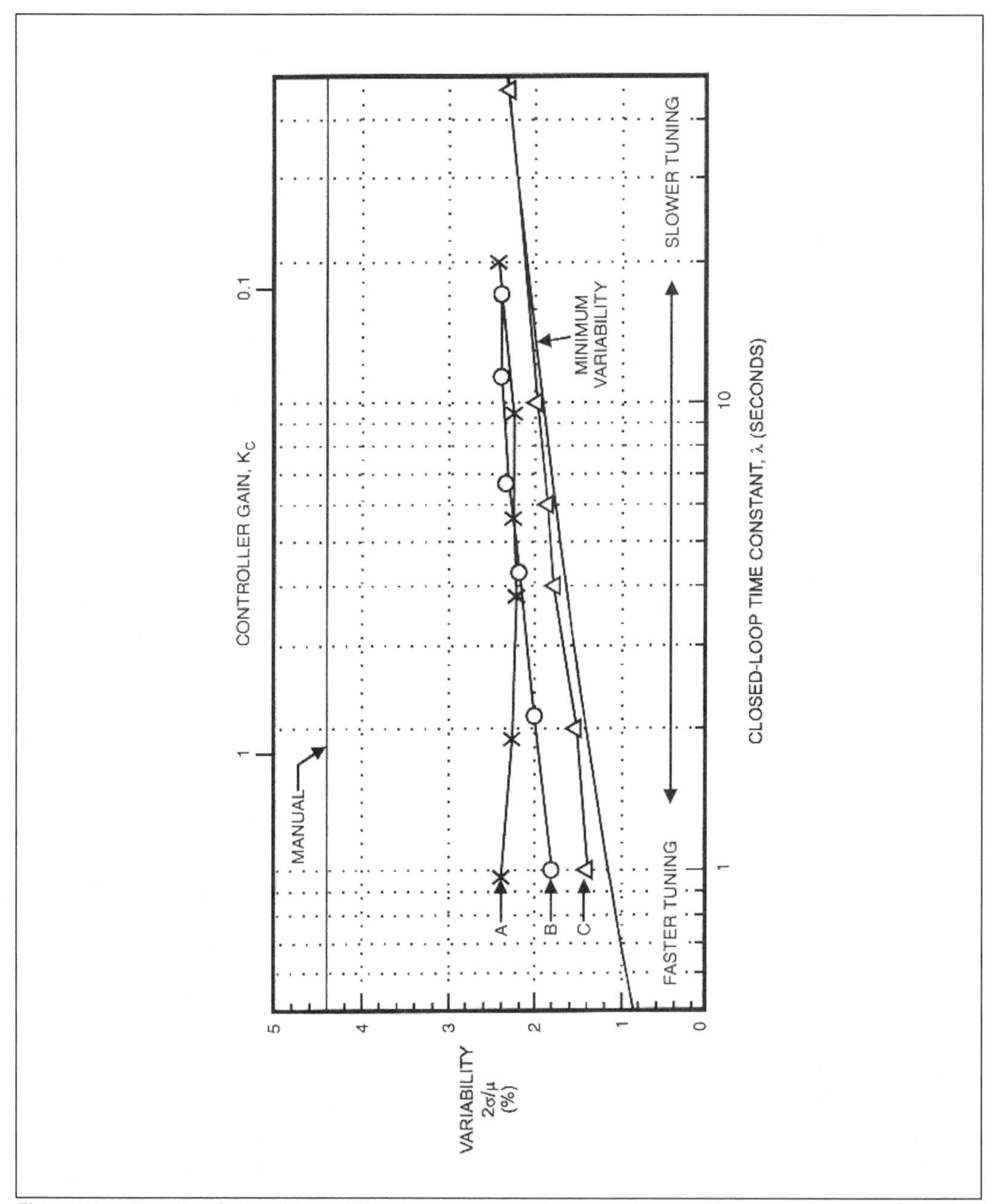

Figure 12-1. Results of closed loop differentiation testing involving three separate valve types showing differences in being able to hold the controller's set point. Note, valve "C" performs closest to an "ideal" curve. (Courtesy of Emerson Process Control)

slow temperature control system, but they start to deviate markedly with increasing speed (shorter time constant such as is found in a pressure or flow control loops). For example, valve "A" exhibits a variability of 2.4% for a time constant of one second, compared to valve "C," which gives much tighter control, having a

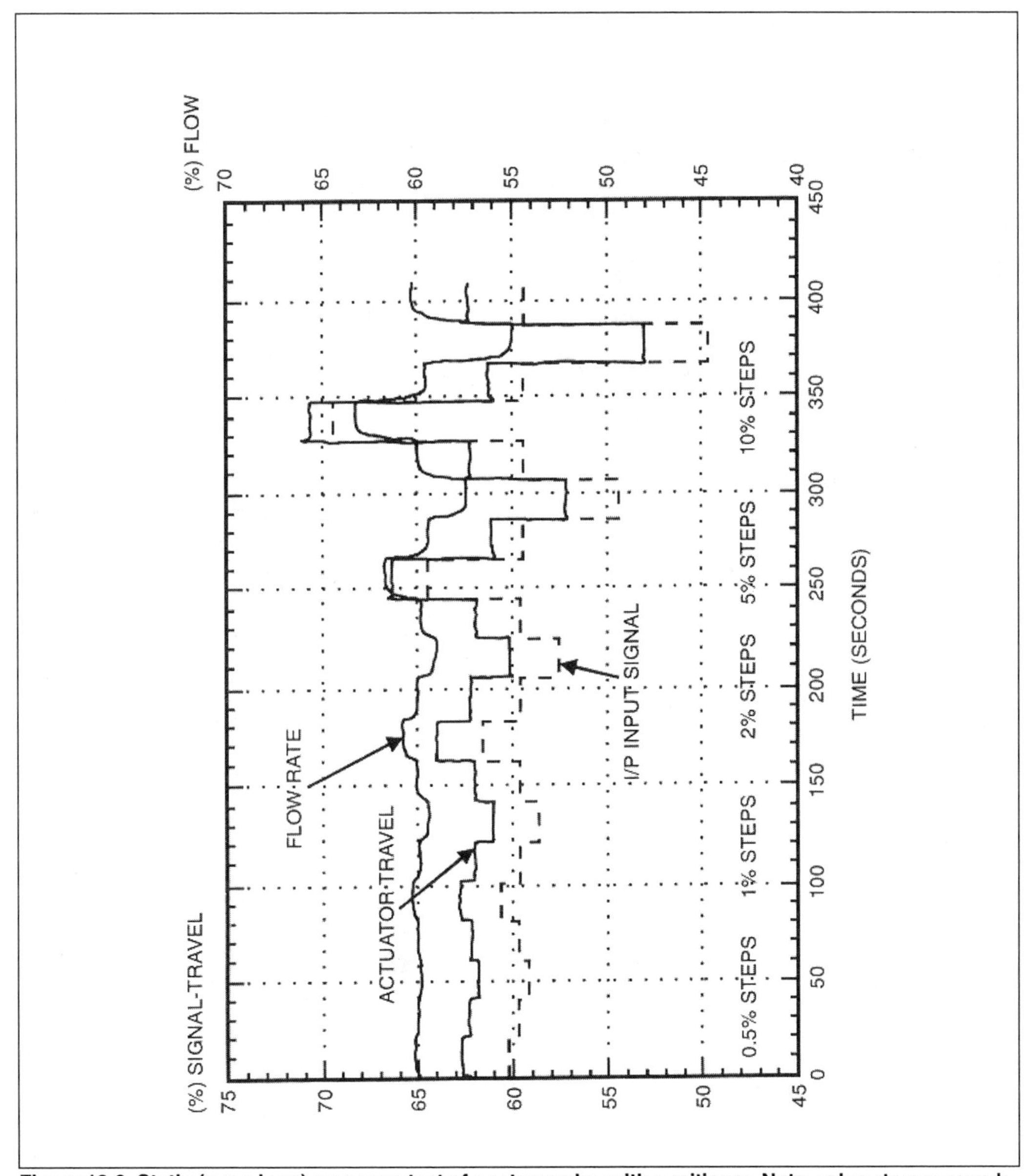

Figure 12-2. Static (open loop) response test of a rotary valve with positioner. Note, valve stem responds to 0.5% signal change but flow does not, indicating lost motion between the stem and closure part. (Courtesy of Emerson Process Control.)

variability of only 1.4%. Valve "B" falls in between. You could not foresee such a difference just looking at the catalog data of each valve!

When looking at the static test data represented by Figure 12-2, you will find that a poor valve/positioner combination sometimes will not even respond to a 1% change in signal (the latter could be an indication of too much stem friction). Dynamic influences causing changes of valve stem forces, due to dynamic instability at certain travel positions, also show up and can be diagnosed.

To achieve good control, we would normally expect a control valve assembly to be able to respond to process variable changes of 1% or less. Note in Figure 12-2 that this valve is capable of accurately producing a travel change for as little as 0.5% change in input signal. However, there is no observable flow change until the signal exchange reaches 1%. This is an indication of a loose connection between the actuator and the closure member. This could happen with some rotary valves where there is clearance in the connection between the actuator stem and the valve stem!

When making these kinds of tests, it is important to compare the flow change in relation to the input signal rather than comparing the stem travel to the input. A stem position change does nothing to correct the process variable unless a corresponding flow change occurs.

Finally, such a test system can measure the installed gain (installed flow characteristic slope) of a valve and the minimum and maximum rate of flow (say in gallons per minute) that can be controlled by a given valve in the test loop. While the "process gain" is typically 1 with the proper transmitter settings, the installed gain of the valve is usually higher. It is important that the "average gain" of the valve (typically between 1.2 to 2.5)* does not deviate by a ratio of more than, say, 2:1. For example, if the average installed gain of the valve is 1.5 and if at 8% of rated flow the gain is 3 going to 3.2 at 7% flow, and at 90% of rated flow the measured gain is only 0.75 (3 being the upper limit and 0.75 the lower limit of the gain), then we could say that the "controllable flow range" is 90%/8% = 11.25:1. This indeed would be a good way to define "installed rangeability."

Perhaps, not surprisingly, this "differentiation testing"[2] proved that well-designed rotary control valves (assuming no "play" between stem and actuator, or, closure member) perform equally as well as good globe valves and contrary to our instincts, a rugged, heavy-duty globe valve does not necessarily outperform a well-designed smaller and lighter-duty valve when it comes to meeting diverse loop conditions. This does not mean that the small valve can handle high pressure drop applications, for example. Loop behavior, after all, is only one part of the valve evaluation process.

---

* Installed control valve gain is defined as the ratio of flow change per controller signal change. For example, a valve having a flow change of 10% of maximum flow for a signal change of 5% has an installed gain of 2. Many people mistakenly believe that the "installed gain" has to be one, i.e., 10% flow change per 10% signal change. This cannot be since it would require 100% valve travel at rated flow (This violates the 80% rule, see Chapter 5, subsection on "What size valve to choose"). In any case, whatever the gain is, the "proportional band," or gain of the controller, can easily be adjusted to match the gain of the valve as long as the latter is relatively constant throughout the applicable flow rates.

REFERENCES

1. Taylor, C., Davey, B., "Control Valve Performance Requirements for Optimum Process Control," *Proceedings of the 34th Annual Int'l Pulp & Paper Symposium*, ISA, (1994) pp. 5-1-5-7.

2. ISA-75.25.01, *Test Procedure for Control Valve Response Measurement from Step Inputs*. ISA, Research Triangle Park, NC: *2000.*

3. ISA-75.13, *Method for the Evaluating the Performance of Positioners with Analog Input Signals and Pneumatic Output*. ISA, Research Triangle Park, NC, 1996.

4. ISA-75.25.02, ISA, Research Triangle Park, NC.

5. Ali, R., "Smart Positioners to the Rescue," *InTech*, June 2002. pp. 72-73.

13

WHEN DO I NEED TO "HARD FACE" THE VALVE TRIM AND OTHER QUESTIONS CONCERNING VALVE MATERIAL

Besides corrosion, which can be handled by consulting a good handbook when selecting a suitable material for wetted parts, erosion is the cause of most valve failures. We need to distinguish the causes of erosion in order to affect a durable solution.

EROSION CAUSED BY SOLID PARTICLES IN THE FLUID STREAM OR GRANULAR MATERIALS SUCH AS COAL SLURRY

If the velocity of the erosive fluid is low enough (say, less than corresponding to a 10 psi valve pressure drop), then elastomer lining works very well, particularly with natural rubber or GRS (BUNA-S) types of elastomers, and a pinch valve or rubber-lined butterfly valve should be considered.

Remember, you need a pressure drop across the valve of at least one psi to keep particles in suspension (3 to 7 feet/second velocity). For higher pressure drop or where temperature is elevated, consider angle valves in the "flow tending-to-close the plug" direction. Try to discharge the flow into a downstream pipe that has a diameter of at least four times the orifice size in order to avoid cutting through the pipe wall. Hard facing of the seating surfaces, the seat ring bore, and the outer contour of the plug is a good practice to prolong service life. For tight shutoff, it is best if there is a difference in hardness between the plug (40 to 46 RC*) and seat ring (56 to 62 RC). The same can be said for the hardness difference

---

* Rockwell C-Scale Hardness

between plug guide and guide bushing. For smaller size plugs, solid ceramic (ALO_2), or carbides[2] (such as tungsten carbide) have been found to be effective. However, the key here is to streamline the fluid flow in all areas that approach the plug/seat restriction (vena contracta) and to avoid anything that would cause direct fluid impact on adjoining valve housing surfaces. Ideally, the valve should only accelerate the fluid.

The actual deceleration (pressure drop) with the resultant, very destructive, high turbulence should take place inside a large pipe or vessel downstream – but not close to the valve plug.

As an example of how brief the service life of valve plugs and seat rings can be, I would like to cite Reference 3, in which a 6 in. seat ring made from 316 stainless steel was completely destroyed within a few weeks while handling fine clay in suspension in hot oil at a pressure drop of only 50 psi (I suspect that the valve also underwent cavitation, which, of course, accelerated the destruction). In another example, a fully Stellite® No. 1-coated 1 1/2 in. diameter valve plug was completely worn away by uranium ore slurry within two weeks even though the pressure drop was only 100 psi.

Remember, the hardened and expensive valve trim will only delay the destruction, not prevent it. In some cases it has been found to be more cost effective to buy the cheapest valve possible and to replace it on a regular basis. This has been the practice with the control of bauxite slurry where a simple carbon steel butterfly valve was used and periodically replaced.

CAVITATION

This process can be avoided in most cases by selecting a high F_L type trim and by perhaps relocating the valve to a point of higher static pressure, by placing an additional restrictor downstream, or by injecting air or gas into the valve near the plug or vane (see Figure 14-7). Selecting a hardened trim will only prolong the service and, of course, will increase the revenue of your vendor from the sale of spare parts (see also Chapter 14). Some cavitation, especially with hot water (condensate) is due to placing the valve too far away from the flash tank.

EROSION BY WET STEAM

On one side of the temperature scale is water condensate, which attacks carbon steel. So only use chrome-moly alloys, bronze, or stainless steel as valve body material, or make sure a gasket is placed under the stainless steel seat ring in case a carbon steel body is employed. On the other side is superheated steam, which can discharge water droplets during isentropic throttling in a valve. Here the use of martensitic, chromium stainless steel trim is recommended, especially AISI Type 440C, which can be hardened to about 58 RC. For lower temperatures (<750°F), 17-4 pH stainless steel is also useful if heat treated, usually at 1075°F, yielding a hardness of about 38 RC. Watch the pH level of the steam. A caustic condition (>7 pH) can cause stress corrosion on parts harder than 35 Rc! Incidentally, this also applies to "sour" (sulfur-containing) natural gas.

There is no hard and fast rule for when or under what pressure conditions a trim should be hard faced. Remember, gaseous fluid undergoes sonic velocity in the orifice (typically 1350 feet/second for steam) regardless of whether the pressure drop is from 3000 psi to 500 psi or only from 15 psi gage to atmosphere.

The reasons for trim wear are more subtle. Is the steam handled by the valve fairly wet, because the valve is far removed from the boiler? Does the fluid contain a solid catalyst? Does the fluid tend to crystallize under certain conditions? Can two-phase flow (liquid plus gas) exist? Does natural gas coming from a well contain sand? These are all questions that should be asked when specifying the valve material. Otherwise, the general-purpose trim material, i.e., AISI type 316 stainless steel, will do just fine with clean liquid or gases even if the pressure drop exceeds 1000 psi, even though most handbooks limit the allowable ΔP for stainless steel trim to between 150 and 200 psi.[3]

A final word regarding guide material. For globe valves and low pressure service (<500 psi ΔP), both plug or stem guide bushing can be 316 stainless steel, as long as the 316 stainless steel plug surface or the 316 stainless steel stem surface is somewhat work hardened by suitable machining or roller burnishing techniques. Heavier valve types use type 440C stainless steel guide bushings or, for corrosive service, bushings made of or faced with a cobalt-chromium-tungsten alloy such as Stellite®.

For rotary valves, TFE-dispersed glass fiber bushings work well up to 350°F and TFE-dispersed bronze bushings to 450°F. Other suitable bearing materials are Nitronic 60®, a stainless steel alloy suitable to 800°F or, if properly retained, a solid graphite bushing. For higher temperature (to 1250°F) and/or corrosive service, use cobalt-chromium-tungsten alloy or graphite.

To recommend suitable materials for the infinite variety of corrosive chemicals is not possible. However, when in doubt, specify a PTFE (Teflon®)-lined valve, which is usually your best bet. Here your choice is between globe style valves with C_v ranges from 0.001 to 100 or rotary valves for larger capacities.

REFERENCES

1. *Compass Corrosion Guide II*, "A Guide To Chemical Resistance of Metals and Engineering Plastics," La Mesa, CA: Compass Publications.

2. Kennicott, W.L., "Use of Hard Carbide Components" In *Petrochemical Processing*, ASME PAPER 61-SA-51 (1962).

3. Wing, Paul, "Selecting Control Valve Trim For High Pressure Drop And Erosive Service," *ISA Journal*, Vol. 2:332-337, Sept. (1955).

14

CONCERN FOR THE ENVIRONMENT

WILL MY VALVE BE TOO NOISY?

The three basic types of noise sources in valves are: (1) mechanical vibration, (2) turbulence and cavitation with liquids, and (3) aerodynamic noise.

Noise caused by resonant vibration of a valve plug within the radial clearance between plug guide and guide bushing, or between plug skirt and seat ring, is usually the result of severe fluid turbulence and can easily exceed 90 decibels of sound level at frequencies between 0.5 and 7 kHz. This phenomenon seems not so much related to pressure drop (fluid velocity) itself but rather to unstable fluid flow such as flashing or cavitating water. Fortunately, this is a rather rare occurrence. Based on my experience, resonance occurs more often where the undamped natural frequency of the mass m (weight of valve plug divided by the acceleration of gravity) is high in relationship to the stiffness K_n (the perpendicular deflection rate) of the valve stem. This happens typically where the undamped natural frequency f_n is below 5000 Hz (cycles/sec.). Now $f_n = 0.16\ (K_n/m)^{1/2}$ in cycles/second. A hollow plug will reduce the mass m while a shorter unsupported stem length, or a larger stem diameter, will increase K_n – *all* positive remedies. Decreasing the guide clearance will reduce the amplitude but is no sure cure of the problem that, when unchecked, can lead to fatigue breakage of stem or plug guide. Fortunately one can hear the resonant vibrations of a valve plug occurring at frequencies mostly between 100 and 1500 Hz and sound levels that can exceed 90 DBA. This then should be taken as a call for remedial action.

The OSHA limit of 8 hours of exposure for workers at 90 dB is measured on the A-weighted scale, which is patterned after the frequency response curve of the human ear; hence, limit = 90 dBA. This is one criterion. On the other hand, noise is nothing but audible vibration, and vibration can cause damage to valves, piping and accessories.

Even if you ignore the OSHA limit by putting the valve underground, the pipe will typically break at a sound pressure above 130 dBA[1] measured one meter from the pipe wall. Therefore, tolerate no more than 110 dBA without using "low noise trim" or other *internal* noise abatement means even when the valve is buried!

Don't read on if your valve, handling gas or steam, has an energy coefficient (EC) of less than 1000. This number is derived by multiplying your absolute inlet pressure in psia by the calculated valve C_v and F_L numbers. For example, you reduce 150 psia steam and the required valve $C_v = 125$ at an F_L factor of 0.9. Here your EC = 150 x 125 x 0.9 = 16875 – potentially a noisy standard valve. On the other hand, a valve with a C_v of only 6 has an EC of 6 x 150 x 0.9 = 810 and, therefore, a sound level of less than the OSHA limit of 90 dBA, regardless of pressure drop.

ISA published a reasonably accurate prediction method in ISA-75.17, Control Valve Aerodynamic Noise Prediction, which was based on modified free jet noise theories.[2] This standard has now been replaced by an international standard IEC 60534-8-3[3] adopting most of the methodology of ISA-75.17 standard.

The 46 plus equations should not scare you if you possess a full size computer, besides the vast majority of vendors have computer programs in place or available to you. However, to make life easier, I have condensed the document to a handier level and which can be used with the aid of a hand-held calculator,

where:

$$L_a = A + L_{pi} - T_L + L_G \text{ dBA} \tag{14-1}$$

$$L_{pi} = 147 + \eta m + 10 \log (C_v F_L P_1 P_2 / D_i^2) + 10 \log (C_o / 1128) \, dB \tag{14-2}$$

ηm = a modified (not the real) acoustical efficiency factor[3] (remember this is a short cut);

P = pressure, psia; (P_1 = inlet; P_2= outlet pressure)

D_i = the inside diameter of the outlet pipe in inches; and

C_o = speed of sound of fluid, ft/sec. (typical numbers: air is 1125 ft/sec, sup. heated steam is 1350 ft/sec).

k = specific heat ratio (1.4 air, 1.33 steam)

T = fluid temperature, °F

M = molecular weight (29 air, 18 steam)

A = $5 - 10 \log (f_o / 6000)$ in dB, where $f_o = f_r \times C_{o\,gas} / 4\, C_{o\,air}$; $f_r = 62420 / D_i$

$$C_o = 223 \sqrt{\frac{k(T+460)}{M}}; \text{ ft/sec.}$$

For example: C_o air = 1128 ft/sec. C_o Helium = 3188 ft/sec. at 70°F

If P_1/P_2 is less than P_1/P_2 *critical*, i.e., subsonic (see Table 14-1).

This is called Regime I, i.e., $P_1/P_2 < P_1/P_2$ *critical,* here

$$\eta m = 26 \log [(P_1 - P_2)/(P_1 F_L^2 - P_1 + P_2)] + 10 \log (F_L^2) - 38.6 \text{ dB} \qquad (14\text{-}3)$$

For Regimes II and III, i.e., $P_1/P_2 > P_1/P_2$ *critical,* flow is sonic up to Mach 1.4:

$$\eta m = -40 + 37 \log [(P_1/P_2)/(P_1/P_2 \text{ critical})] + 10 \log (F_L^2) \text{ dB} \qquad (14\text{-}4)$$

For Regime IV, i.e., $P_1/P_2 > P_1/P_2$ break, jet velocity is above Mach 1.4:

$$\eta m = -31.2 + 10 \log (F_L^2 + 10 \log [(P_1/P_2)/(P_1/P_2 \text{ break})] \text{ dB} \qquad (14\text{-}5)$$

For Regime V, i.e., $P_1/P_2 > 22\alpha$, jet velocity is above 3 and acoustical efficiency is constant:

$$\eta m = \eta m \text{ Max. of Regime IV} \qquad (14\text{-}6)$$

NOTE: P_1/P_2 *critical* $= P_1/(P_1 - 0.47 F_L^2 P_1)$. P_1= inlet pressure, psia; P_2= outlet pressure, psia.

Now the transmission loss is:

$$T_L = T_{Lfo} + \Delta T_{Lfp} \quad dB \qquad (14\text{-}7)$$

where:

$$T_{Lfo} = 10 \log[9 \times 10^6((P1/P2)+1)\, r\, t_p^2/D_i^3] \quad dB \qquad (14\text{-}8)$$

D_i = pipe inside diameter, inches
G_g = specific gravity (air = 1, saturated steam = 0.49)
r = distance from the pipe = 39 in. + $1/2\ D_i$
t_p = pipe wall thickness, inches
C_o = Speed of sound in fluid; air = 1128 ft/sec, steam = 1320 ft/sec.

and

T_{Lfp};

$$\text{If } f_p \le f_o, \text{ then } \Delta T_{Lfpo} = 20 \log(f_o/f_p) \qquad (14\text{-}9)$$

$$f_p \le 4 f_o, \text{ then } \Delta T_{Lfp} = 13 \log(f_p/f_o) \qquad (14\text{-}10)$$

$$\text{If } f_p > 4 f_o, \text{ then } \Delta T_{Lfp} = 20 \log(f_p/f_o) + 7.8 \qquad (14\text{-}11)$$

$$If\ f_p > 4\,f_r\ then\ T_{Lfp} = 12 + 10\log(f_p / 4f_r) \tag{14-11a}$$

where:

$f_r = 62420 / D_i$, cps

$$f_o = 0.25\,f_r,\ \text{based on air, cps} \tag{14-12}$$

$$f_p = 0.2\,C_o \times M_j / D_j;\ if\ P_1/P_2 < 3,\ \text{cps} \tag{14-13}$$

$$f_p = 0.28\,C_o / D_j\,(M_j^2 - 1)^{0.5},\ \text{above}\ P_1/P_2 = 3 \tag{14-14}$$

The Mach number is:

$$M_j = \left\{\left(\frac{2}{k-1}\right)\left[\left(\frac{P_1}{P_2\alpha}\right)^{(k-1)/k} - 1\right]\right\}^{0.5} \tag{14-15}$$

Table 14-1. Important Pressure Ratios *(P₁/P₂)* for Air

F_L	0.5	0.6	0.7	0.8	0.9	1.0
α	0.60	0.63	0.69	0.75	0.85	1.00
P_1/P_2 critical	1.13	1.20	1.30	1.43	1.61	1.89
P_1/P_2 break	1.95	2.07	2.24	2.40	2.76	3.25
P_1/P_{vc} crit. = 22α	13	14	15	16	19	22
P_{vc} crit. = $0.53\,P_1$						

The jet diameter, D_j, is a function of the flow area and the valve style modifier F_d (see Table 14-2); $\alpha = (P_1/P_2\ \text{critical})/1.89$ (see Table 14-1).

If you have a multiple-hole, low-noise trim, then $F_d = 1/(N_o)^{0.5}$, where N_o is number of exposed holes. (4)

$$D_j = 0.015\,F_d\,(C_v\,F_L)^{0.5}\ \text{in feet} \tag{14-16}$$

(See Table 14-2 for representative values of F_d.)

Finally, L_G, the pipe velocity correction factor, is:

$$L_G = 16\log\left[1 \Big/ \left(1 - \frac{0.02 P_1 C_v F_L}{D_i^2 P_2}\right)\right]\ \text{dB} \tag{14-17}$$

NOTE: The above method is limited to valve outlet Mach numbers M_{ni} of 0.3 for standard valves and 0.2 for low noise valves, where

$$M_{ni} = 1 - \frac{0.02 P_1 C_v F_L}{D_i^2 P_2} \qquad (14\text{-}18)$$

Table 14-2. Typical F_d Values of Control Valves, Full Size Trim

Valve Types	Flow Direction	Max. C_v/d^2	F_N	F_d @% of Rated Flow Capacity (C_v)					
				10	20	40	60	80	100
Globe, parabolic plug	To Open	13	0.28	0.10	0.15	0.25	0.31	0.39	0.46
Globe, 3 V-port plug	To Open	10	0.44	0.29	0.40	0.42	0.43	0.45	0.48
Globe, 4 V-port cage	Either*	10	0.38	0.25	0.35	0.36	0.37	0.39	0.41
Globe, 6 V-port cage	Either*	10	0.26	0.17	0.23	0.24	0.26	0.28	0.30
Globe, 60-hole drilled cage	Either*	8	0.14	0.40	0.29	0.20	0.17	0.14	0.13
Globe, 120-hole drilled cage	Either*	6.5	0.09	0.29	0.20	0.14	0.12	0.10	0.09
Butterfly, swing-through, to *70°*	Either	32	0.34	0.26	0.34	0.42	0.50	0.53	0.57
Butterfly, fluted vane, to 70°	Either	26	0.13	0.08	0.10	0.15	0.20	0.24	0.30
Eccentric rotary plug valve	Either	13	0.26	0.12	0.18	0.22	0.30	0.36	0.42
Segmented V-ball valve**	Either	21	0.67	0.60	0.65	0.70	0.75	0.78	0.80

Notes: *Limited $P_1 - P_2$ in flow towards center
d = valve size in inches
F_N = valve specific noise parameter, defined by the author as the F_d at a flow capacity (C_v) equivalent to 6.5 d^2. Depending upon pipe size, a lower F_N means less external noise due to higher pipe wall attenuation.
**For sites 6" and above.

To estimate sound levels at pipe velocities higher than the given limits, consult Reference 10.

If this is still too complicated for you, then here is my second method, as simple as A(+)B(+)C.

First establish "A." This is the basic noise level of a single-seated globe valve, an eccentric rotary plug valve, a butterfly valve, or any cage valve with no more than four ports per cage, measured with a schedule 40 pipe, 100 psia inlet pressure, and at an opening equal to about 70% of rated C_v shown in Figure 14-1, where the assumed rated C_v number is equivalent to $10D^2$ (i.e., a 2 in. valve having a C_v of 40). Note that Curve "A" by itself will cover probably 50% of all noise evaluations. The higher pressure ratio lines are limited on purpose. Beyond the P_1/P_2 ratios shown, you will see more than 0.3 Mach pipe velocities and the longhand method should be used.

Next, we add "B," a correction for inlet pressure if it is different than approximately 100 psia. B = 12 log $(P_1/100)$ dBA. Thus, for 540 psia inlet, B = 12 log (540/

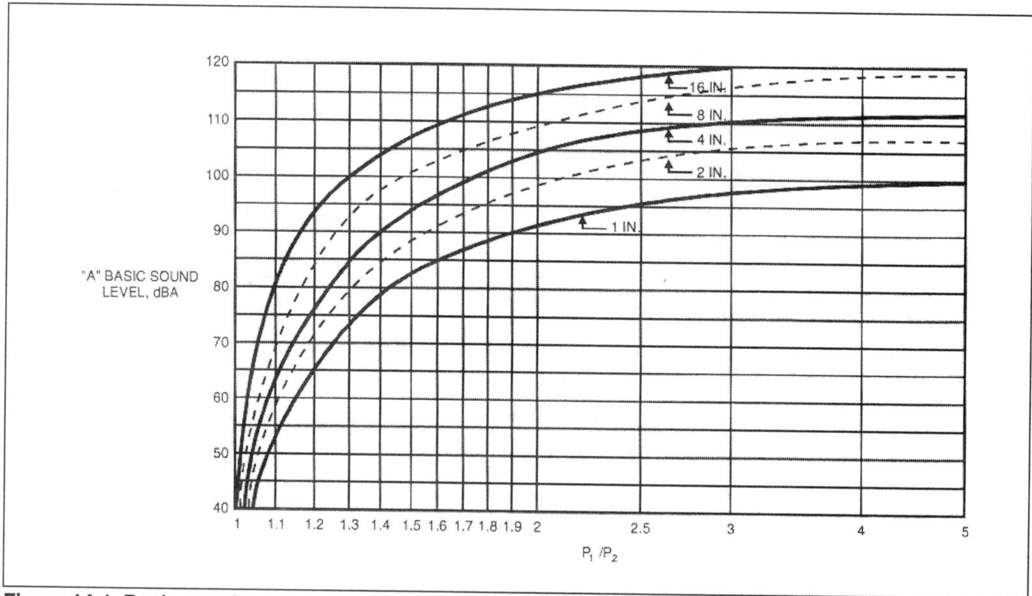

Figure 14-1. Basic aerodynamic sound level in dBA for conventional control valves at approximately 70% of rated flow capacities, at 100 psia inlet pressure, and schedule 40 downstream pipe, measured one meter (3 feet) from the downstream pipe wall.

100) = 8.8 dBA. Finally, add "C," the correction for pipe wall other than schedule 40 (see Table 14-3).

Most gases produce sound levels very close to air. However, you may want to subtract 3 dBA from the total in case of steam. Using as an example, a 3 in. globe valve with a parabolic plug reducing 540 psia steam to 315 psi a in a 3 in. schedule 80 pipe, the C_v calculated is 65 at an F_L number of 0.87. P_1/P_2 = 540/315 = 1.7. From Curve "A" we found no listing for 3 in., but let's read between the lines. At P_1/P_2 of 1.7, we estimate "A" as 97 dBA. Next "B" = 12 log (540/100) = 8.8 dBA. Finally, we add "c" = -4 dBA. The total = 97 + 8.8 – 4 = 101.8 dBA. Subtracting 3 dB for steam gives us about 98.8 dBA at 1 meter from the downstream pipe.

Table 14-3. Pipe Wall Modifier "C"

Pipe Schedule:	20	40	80	160
C, dBA =	+4	0	-4	-8

METHODS OF AERODYNAMIC VALVE NOISE REDUCTION

Now to the more common occurrence of "noise pollution," i.e., aerodynamic noise; from the foregoing equations for the estimation of throttling noise levels, it becomes apparent that two basic variables are mainly responsible for the sound level and, therefore, can be varied to achieve a desired noise reduction. They are the jet velocity U_{vc} and the internal peak noise frequency, f_p.

The jet velocity is the most important element in the noise generation, yet the most difficult one to deal with. Since the velocity is a fraction of the internal orifice pressure ratio P_1/P_0, where P_0 is the static pressure inside the orifice, the manipulation of this ratio is the only way to reduce velocity. Here;

$$P_o = P_1 - \{(P_1 - P_2) / F_L^{0.5}\}$$

In the subsonic Regime I, i.e., $P_1/P_0 \leq 1.9$, selection of a valve trim that is less streamlined, i.e., one that has a high F_L number (0.9 to 1.0) will give a lower jet velocity, hence, lower sound level. At higher pressure ratios the only remedy is to use orifices or resistance paths in series so that, ideally, each sub-jet operates in the subsonic regime. Figure 14-2 shows such a valve that has a multi-step resistance trim (the zig-zag path). Assuming an application where natural gas with an inlet pressure of 1500 psia is to be throttled down to 750 psia, and further assuming the valve pressure recovery factor to be 0.9, the internal pressure ratio (P_1/P_0) for a single-stage reduction would have to be P_1/P_2 = 2.35:1, yielding a modified acoustical efficiency factor of -34.8 dB from Equation 14-4. However, using a trim from Figure 14-2 with 10 reduction steps, the pressure ratio across the orifice for each step now is only 1.089:1 with a modified acoustical efficiency of only –63.1 dB, a 28.3 dB noise saving (ignoring other factors).

Ignoring the partial addition of 10 separate sound power sources, this represents a noise reduction of about 29 dB. (The actual noise savings due to added sources and change in frequency, etc. would be about 20 dB).

A less costly way to reduce "audible" throttling noise is to increase the peak frequency by using a multi-orifice trim, as shown in Figure 14-3. Here, the throttling is still single-stage. Only the jets are divided into a number of parallel ports. This lowers the valve style modifier F_d, since $F_d = 1/(N_o)^{0.5}$, where N_o is the number of equally sized parallel ports. Since F_d is proportional to the jet diameter D_j, the frequency can be shifted to a higher level.

For example, if N_o = 16 as in Figure 14-3, F_d then is 0.25 and the resultant peak frequency is 1/0.25 or four times higher than a single orifice (F_d = 1) with the same total flow area. If the peak frequency occurs in the mass-controlled region of the pipe, which is most likely, an overall reduction in the external (audible) sound level of $\Delta T_{Lfp} = 20 \log (f_{pl}/f_p{}^{0.25}) = 12$ dB is achieved (unless a large pipe is involved, see below). This approach is limited to about 3:1 pressure ratio across the cage wall. At higher ratios, jets may combine downstream and negate the benefits of high frequencies[4]. A note of caution, the above equation only works in regions where the peak internal sound frequency (f_p) is no more than two octaves (4 times) above the ring frequency of the pipe (f_r). The reason is that below this ratio the internal sound shifts from a 20 dB per octave to only 10 dB per octave. This occurs typically when a valve with low noise trim (high f_p) is installed in a large pipe (8′ and larger).

A more elaborate trim arrangement is shown in Figure 14-4. Here the multi-step is combined with the multi-path approach in a layer of disks having individually cast or etched channels. In this design, we have both the beneficial effects of reduction in velocity combined with the benefits of higher transmission loss caused by the increase in frequency. Typical noise reductions are up to 30 dB.

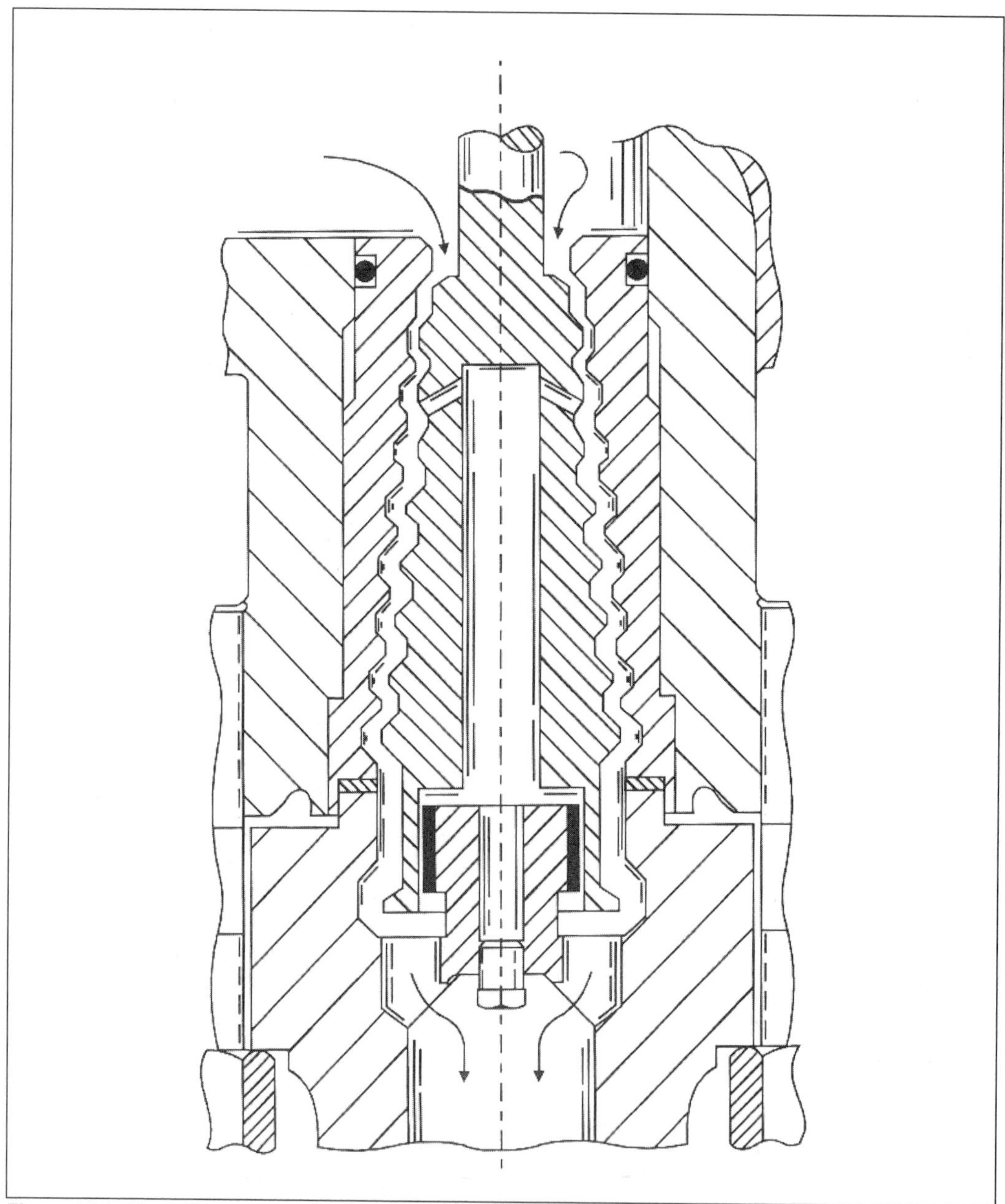

Figure 14-2. Single flow area, multi-step valve plug. Low noise valve trim using 11 throttling steps with single, annular flow area. NOTE: Area enlarges to accommodate change in gas density with lowering of pressure. Benefits are lower throttling velocities.

Note: Pay attention to the valve outlet diameter. An outlet velocity above 0.2 Mach can negate the beneficial effects of this trim due to the resultant turbulence created sound in the downstream pipe.

Figure 14-5 shows a recent improvement over the labyrinth trim. It consists of multiple and identical layers of laser cut plates which, between them, create a three-dimensional flow path typically from a single streamlined flow path on the

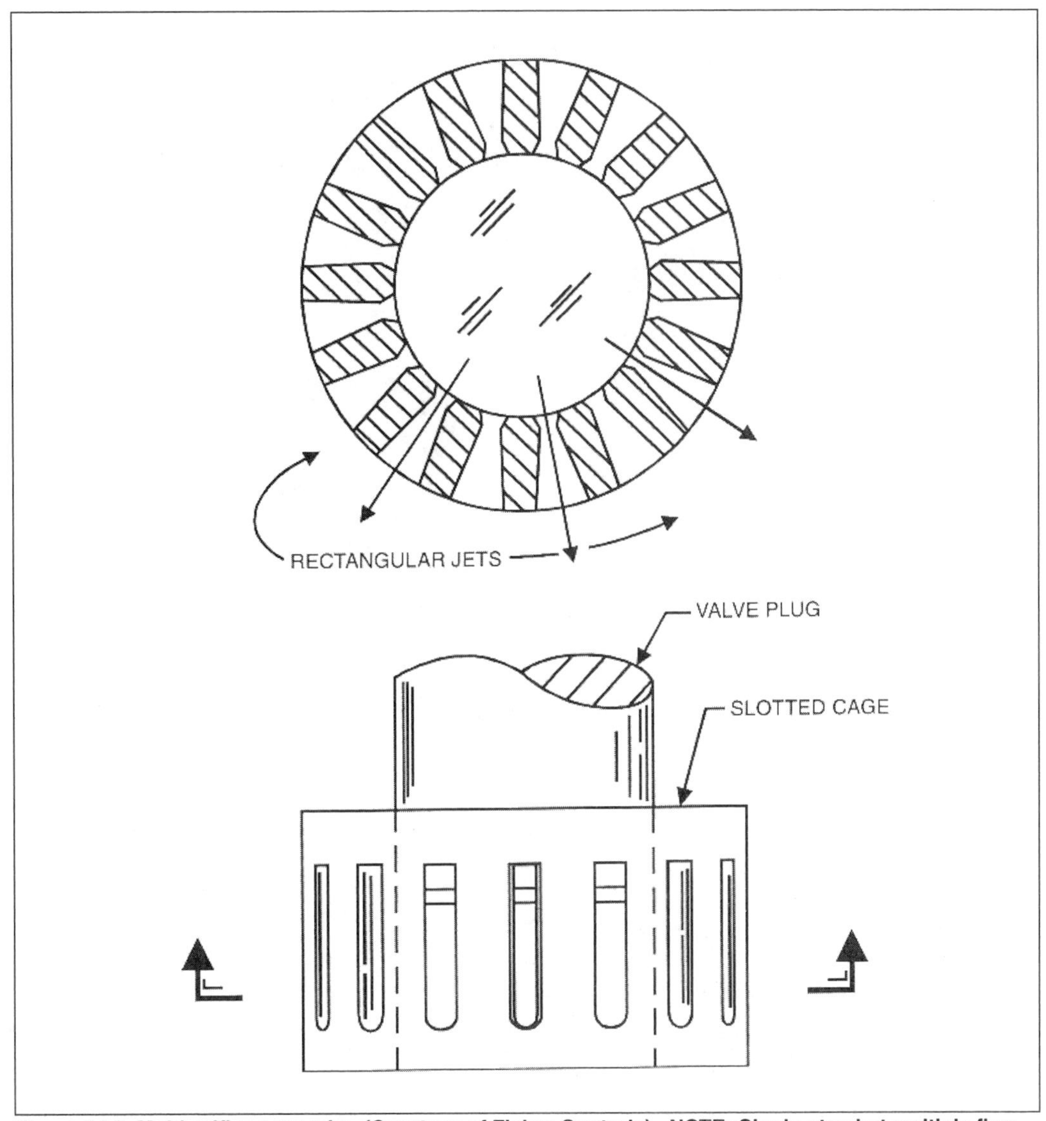

Figure 14-3. Multi-orifice cage trim. (Courtesy of Fisher Controls) . NOTE: Single-step but multiple flow passages characterize this cage trim. Benefits are higher internal sound frequencies and resultant increase in pipe transmission loss.

inside of each disk to multiple non-streamlined rectangular outlet flow passages at the outer circumference. The ratio between the number of outlet to inlet flow passages is dictated by the overall pressure ratio across the valve. In order to obtain maximum noise reduction, the outlet stage is kept subsonic (where a less streamlined path has lower acoustic efficiency), while the first stage sees supersonic flow velocities. Here, the streamlined passages act as supersonic diffusers converting most pressure reduction energy in the form of shockwaves (instead of generally more sound producing turbulence). Also beneficial is the fact that a good portion of the sound energy produced by the first stage is retained in the "settling space," which is the vertical passage between the two stages. For exam-

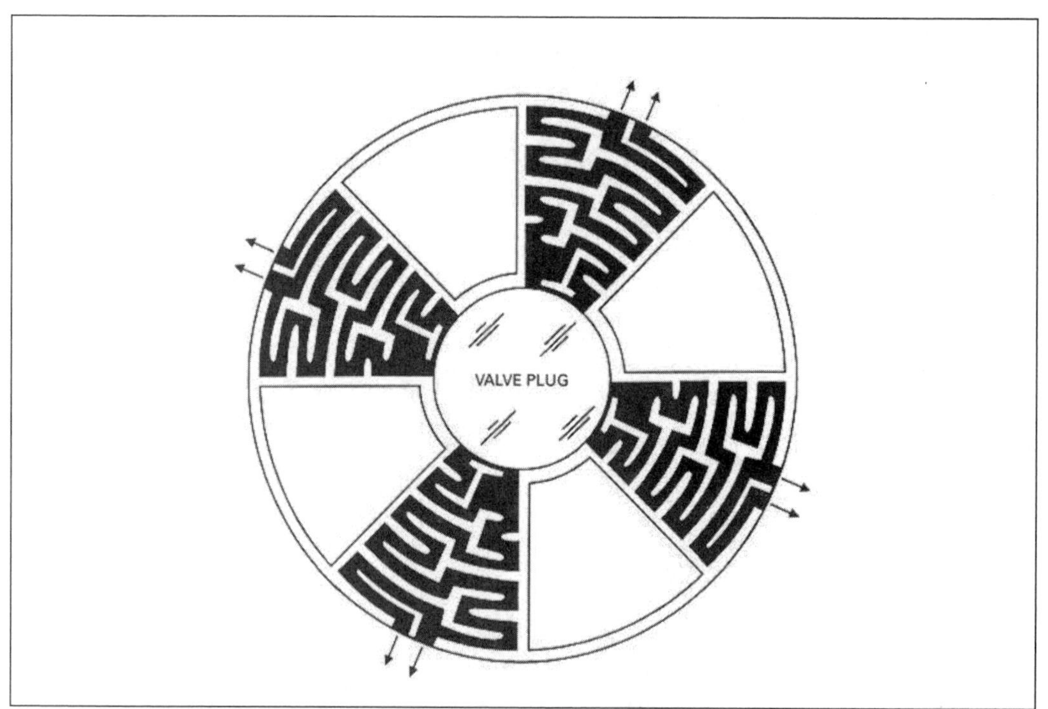

Figure 14-4. Labyrinth-type cage insert. NOTE: Staked washer-like cage trim featuring a combination of multi-step and multi- path grooved flow channels for reduced velocity and increased frequency benefits.

ple, if the overall area ratio between the outlet and inlet passages is 6:1, then the pressure ratio across the first stage is typically 5:1 while the overall pressure ratio across both stages can be as high as 8:1 while maintaining the optimized low noise performance. The absence of a long, torturous path makes this trim much narrower than a typical labyrinth trim (see Figure 14-4); hence, it has more flow capacity for a given trim size. Finally, another benefit is that there is less likelihood of the flow passages plugging up if dust particles are present in the fluid; a problem typically associated with trim sets operating with low gas velocities and fine flow passages.

Finally, a more economical solution is to couple a static downstream pressure reducing device, such as a multiple-hole restrictor, with a throttling valve such as the butterfly valve with a fluted disk shown in Figure 14-6.

Again, the fluted disk will generate multiple, parallel jets with increased noise frequencies similar to the multi-port cage valve. However, most of the pressure energy will be converted by the static plate, resulting in noise reduction at a maximum flow of about 20 dB. At maximum design flow, the valve is sized to have a low pressure ratio of 1.1 or 1.15 where the rest of the pressure reduction occurs across the static plate, which has inherent low noise throttling due to multi-stage, multi-hole design features.

Placing a silencer downstream of a valve is not cost effective owing to limited noise reduction (approximately 10 dB). This is because a good portion of the sound power travels upstream (particularly in so-called "line of sight" valves such as butterfly and ball valves), or radiates through the valve body and actua-

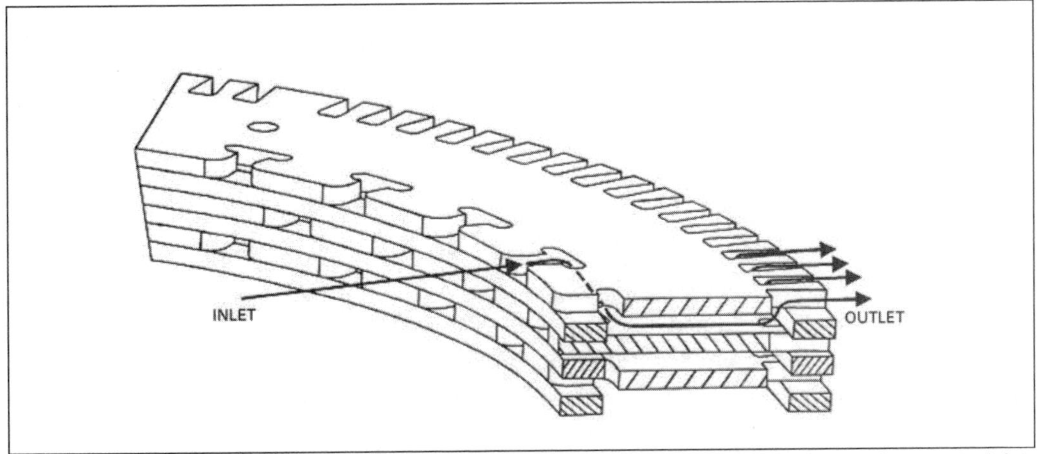

Figure 14-5. Two-stage, multi-area, low noise trim fabricated from identical stampings or laser cuts. Inlet passages have a streamlined configuration to create supersonic expanding jets under high pressure ratios while outlet passages, with increased flow areas, have abrupt passages for subsonic outlet velocity. (Courtesy of Fisher Controls)

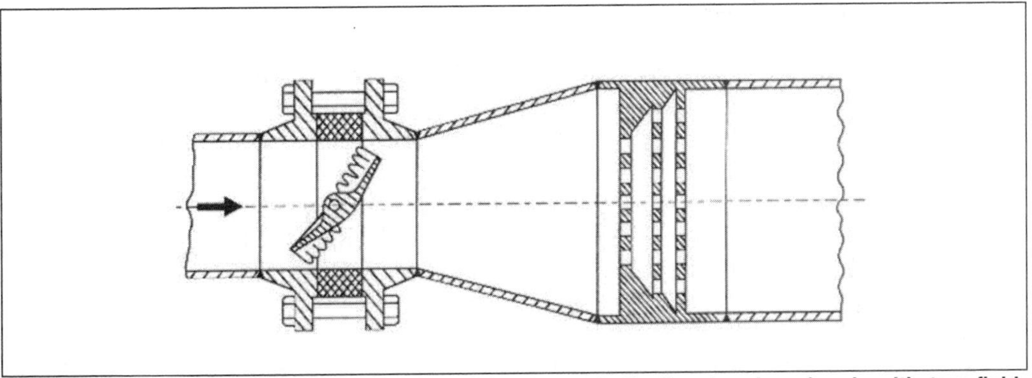

Figure 14-6. Combination of control valve (fluted, low-torque butterfly valve shown) and multi-stage fluid restrictor. (Courtesy of H.D. Baumann Inc.). NOTE: Restrictor is sized to absorb 90% to 95% of available differential pressure at max flow conditions.

tor. Similar concerns apply to acoustic insulation of the pipe wall where the effectiveness, again, is limited to about 15 dB.

The following is a summary of valve noise abatement procedures:

- If $C_v \; x \; F_L \; x \; P_1$ is less than 1000, then no problem [SL $\leq$ 90 dB(A)].
- 1 in. thick pipe insulation = 5 to 10 dB reduction.
- Doubling pipe wall $\cong$ 6 dB reduction.
- Placing silencer downstream $\cong$ 10 dB reduction.
- With silencer up and downstream $\cong$ 20 dB reduction.

- Multi-port resistance plate downstream of valve = 15 to 20 dB reduction. (Use only 5% to 10% of valve inlet pressure as valve ΔP at maximum flow.)
- Special low noise valve = 15 to 30 dB reduction.

Of course, you can apply a combination of the above methods, which is sometimes more cost effective.

A note of caution: Low Noise trims operating in high frequencies typically between 5000 Hz and 20,000 Hz can interfere with ultrasonic flow meters when placed downstream of such valves where high valve trim produced frequencies can reduce the accuracy, or even cause failure of such flow meters.[5]

HOW ABOUT CAVITATION?

I used to say, only half in jest, that cavitation is the manufacturer's best friend, since it supplies a steady demand for replacement trim parts. No wonder, since the mechanism of cavitation (violent collapse of vapor bubbles in a liquid following deceleration of the liquid flow and subsequent pressure recovery) can cause impact pressures on the surface of the valve plug, the inner body wall, or the pipe wall up to 1.5×10^3 N/mm^2 for water [6] (218,000 psi!). No wonder there is no material known to man that will withstand cavitation damage. A harder material (higher elastic stress limit) will last longer than a softer material. For example, here is the relative lifetime of some materials (same material loss under the same cavitation conditions).

Soft Aluminum Alloy	1
Type 1020 Carbon Steel	2.4
Type 304 Stainless Steel	6.5
Hardened 42% Cr, 4% Mo Steel	13

This means specifying hardened Cr-Mo trim instead of stainless steel may increase the lifetime of your trim by perhaps a factor of two (13/6.5), and is probably not worth the effort.

A wiser course is to avoid (a) pressure conditions that cause cavitation or (b) selecting a valve with a less streamlined trim (high F_L). Unfortunately, process conditions can seldom be altered. However, we can help by choosing a better valve location. For example, placing the valve ahead of a long pipe or heat exchanger will increase the static inlet and outlet pressures. So will placing the valve at ground level instead of in an overhead pipe (static head due to elevation will increase inlet and outlet pressures). All this increases the allowable pressure drop while avoiding cavitation problems.

Here are three flow conditions or pressure ratios in relation to cavitation that could concern you:

- Incipient Cavitation. This is the pressure ratio at which the first bubbles are formed (detection by acoustic measurement).

$$X_{Fz}{}^{*} = \Delta P/(P_1 - P_v) \text{ or } \Delta P_{inc\ \text{cavitation}} = X_{Fz}\,(P_1 - P_v) \quad (14\text{-}19)$$

where:

P_1 = inlet pressure (psia)

P_v = vapor pressure of liquid (psia)

X_{Fz} can be as low as 0.10 in some valves, although incipient cavitation can be ignored for all practical purposes (does not produce excessive noise except for very large valves and does not produce sufficient damage).

Now, X_{Fz} can be estimated to a reasonable degree of accuracy using the following equation:

$$X_{Fz} = 0.90 \,/\, [1 + 3.2\, F_d\, (0.85\, C_v \,/\, F_L)^{\,1/2}]^{1/2} \quad (14\text{-}20)$$

For example, a valve having an F_d of 0.46, F_L of 0.88, and a C_v of 200 has an X_{Fz} of 0.194.

- Damage Inducing Cavitation. This is the important ratio we should consider. Exceeding ΔP_{damage} will result in significant reduction of trim life. By perhaps a lucky coincidence, this ratio most often also defines the onset of 85 dBA noise level with a schedule 40 pipe!

$$\Delta P_{damage} = K_D\,(P_1 - P_v) \quad (14\text{-}21)$$

As an approximation, and for conventional globe and rotary control valves (standard trim):

$$K_D = X_{Fz}\,(1 + K_S \times K_P \times F_L{}^2) \quad (14\text{-}22)$$

K_S = size coefficient $(1/d)^{0.125}$; d = valve size in inches

Example: An 8 in. valve has a K_S of $(1/8)^{0.125} = 0.77$

K_p is a pressure correction factor, where $K_p = (100/P_1)^{0.125}$

For example, if P_1 = 225 psia, $K_p = (100/225)^{0.125} = 0.90$

---

* See IEC standard 60534-8-4, the incipient cavitation index is also sometimes expressed as σ_i where $\sigma_i = 1/X_{Fz}$.

Assuming further that the 8 inch valve is a butterfly valve, which at an assumed C_v of 640 has an F_L value of 0.75, and an X_{Fz} of 0.16, the coefficient of damage K_D can now be calculated as:

$K_D = 0.16 \times \{1 + 0.77 \times 0.90 \times 0.75^2) = 0.22$ from Equation (14-22)

Assuming the fluid is cold water at P_v = 0.5 psia, the maximum allowable pressure drop ΔP to avoid noise and cavitation damage is:

$\Delta P_d = K_D (P_1 - P_v) = 0.22\ (225 - 0.5) = 49.4$ psi.

Another example: We want to control 180 gpm of a liquid with a vapor pressure at 180°F of 13.5 psia. The inlet pressure P_1 = 125 psia and P_2 = 38 psia. From our C_v calculation, and assuming F_L = 0.9 (G_f= 1), we find the flow is not critical and C_v required = 19.3. Selecting a 1 1/2 in. globe valve, we find the K_D factor from the equation (14-22) to be 0.54

$\Delta P_{damage} = 0.54\ (P_1 - P_v) = 0.54\ (125 - 13.5) = 60.2$ psi

Now the actual ΔP = 87 psi, so we definitely have a damage and a noise problem that calls for an anti-cavitation trim or placement of a resistance plate downstream of the valve (see Figure 14-6). Let's assume we chose the latter. This increases the outlet pressure at maximum flow to 115 psia, and it decreases the valve ΔP to 10 psi, i.e., less than ΔP_{damage}. The C_v requirement of the valve will increase to 57 and you will have to choose a 2 in. size valve. Now what will happen at half the flow? Since the restrictor has a fixed C_v, the ΔP of the restrictor decreases as $(Q1/Q2)^2$, or by a factor of four. This makes the valve $P_2 = 38 + [(115 - 38)/4] = 57.3$. The valve C_v is now 90 gpm/ $\sqrt{125 - 57.3}$, which is 11 since F_L is still 0.9. The corresponding 2 inch valve. K_D factor is 0.97 and ΔP damage = 0.97 (125 – 13.5) = 108.2. This is above the limit of the actual ΔP = 125 – 57.3 = 67.7 psi, illustrating that fixed downstream restrictions work[7] if the flow does not vary by more than 2:1!

- Choked Cavitation Flow. The final cavitation pressure ratio $K_{choked} = F_L^2$. This is the "choked" pressure ratio at which point the flow remains constant with further decrease in P_2;, $\Delta P_{choked} = F_L^2\ (P_1 - P_v)$.

Here the cavitation noise is near its maximum and severe cavitation damage can occur in the downstream piping of rotary valves, for example (also called "super cavitation"). This condition should definitely be avoided except for some short duration emergency conditions.

WHAT TO DO ABOUT CAVITATION

As I mentioned before, try to avoid exceeding ΔP_{damage}. If changing the process conditions is not possible, select an anti-cavitation trim from your vendor's catalog.

Another solution, previously outlined is the application of a single- or multi-stage resistance plate downstream of your reducing valve. The key to success is proper valve sizing. At maximum flow rate, allocate only 10% of the total pressure drop across the valve and 90% across the plate!

A third solution is to place two identical valves in series. This has the effect of having a two-stage trim. The F_L or K_D of the installation is now $\sqrt{F_L}$ or $\sqrt{K_D}$; that is, if a given valve F_L was 0.7, the combined F_L is now $\sqrt{0.7} = 0.84$, thereby giving you a corresponding increase in ΔP capability.

Surprisingly as it sounds, noise reducing trims designed for gases work equally as well for liquids. Most effective are trims having multi-stages with gradually enlarging flow areas. However, instead of allowing for the density changes due to lowered pressures with gases, the enlarged flow areas will slow down the velocity of liquids, thereby creating only a small pressure drop at the outlet. A note of caution regarding stacked labyrinth trims (see Figure 14-4). Here the normal flow direction is from the plug outwards. This works well for gases but will lead to cavitation damage of the plug surface with high pressure liquids (there exists max. pressure drop through the gap between plug surface and the cage bore). Reversing the flow will help, although this requires a reconfiguration of the labyrinth passages (growing larger towards to inlet) with the resultant reduction in flow capacity (C_v).

Lastly, cavitation damage and noise can be reduced by gas or air injection, as shown in Figure 14-7. This simple method works as long as P_v is below 7 psia and the fluid can tolerate gas or air entrainment.

A typical application for this method is backpressure control of water from a cooling tower. By placing the point of air injection closely downstream of the point of jet constriction, the differential between the near absolute vacuum (cold water $P_v = 0.5$ psia) and the 14.7 psia atmospheric outlet pressure will force the air through the check valve and into the fluid to act as a cushion for the imploding vapor bubbles. Since the process takes place downstream, there is no effect on the flow capacity of the valve.

ESTIMATING HYDRO-DYNAMIC NOISE

Here we have to distinguish between noise produced by turbulent flow (only important if large pipes are involved) and noise produced by cavitation. The latter is more common and can create sound levels in excess of 90 dBA even with valves as small as one inch.

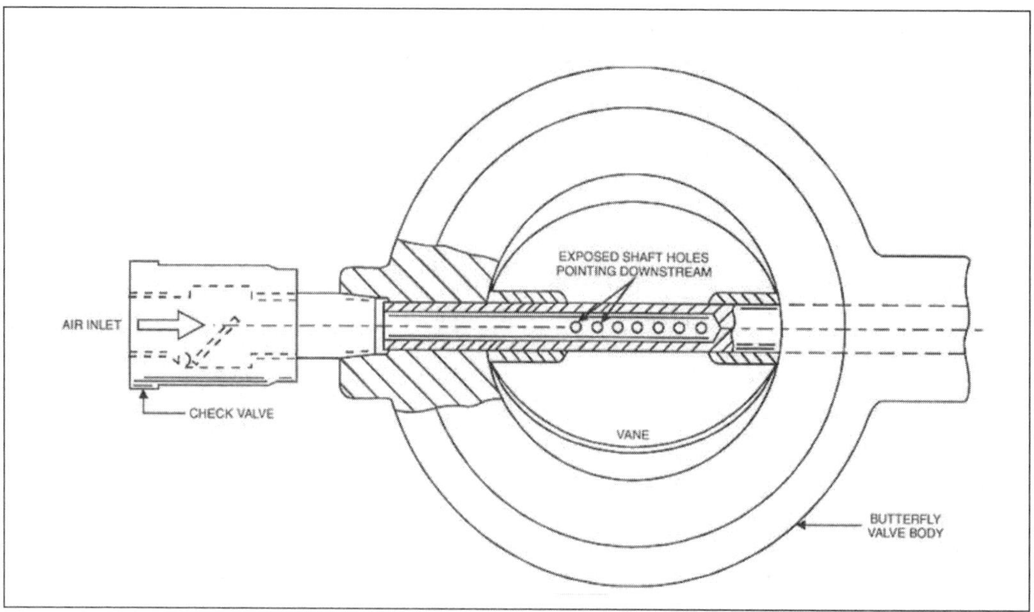

Figure 14-7. Air injection-type butterfly valve to reduce cavitation noise and damage. (Courtesy of Baumann Inc.) NOTE: Air enters through the check valve on left due to cavitation-caused vacuum downstream of the vane[11].

NOISE PRODUCED BY TURBULENCE

The estimated sound level in dBA given by Reference 8 is:

$$Lp_a = 2 + Lp_i + TL_{fptrb\cdot} - 10 \log (D_i + 2\, t_p + 2 \,/\, D_i + 2\, t_p). \qquad \text{dBA} \qquad (14\text{-}23)$$

where $Lp_i = 10 \log (3.2 \times 10^9\, Wa\, \rho_L\, c_L \,/ D_i^2)$ dB.

$$Wa = 0.25 \times 10^{-4}(U_{vc}/c_L)\,(m\, U_{vc}^2\, F_L^2 \,/2) \qquad \text{in Watts} \qquad (14\text{-}24)$$

And where U_{vc} = jet velocity in m/s; $U_{vc} = (1/F_L^2)(2\Delta P_C \,/ \rho_L)^{1/2}$

D_i = pipe 1.DIA, m, t_p = pipe thickness, m, m = flow rate in kg/sec.

In contrast to aerodynamic noise, the point of lowest transmission loss for water at the pipe is at the ring

$$\text{frequency } f_r, \text{ where } T_{Lfr} = -10 - 10 \log(c_p \rho_p t_p / c_o\, \rho_o D_i) \quad \text{dB} \qquad (14\text{-}25)$$

Here P_C is valve pressure drop in Pascal and ρ_L the density of liquid in kg/m^3 (water is 1000). For other densities; subscripts p is for pipe material and o for air. C = speed of sound, subscript o is for air and p for pipe material. t_p is the pipe wall thickness. Unfortunately, there is additional change in transmission loss due to a change in the peak frequency, a rather complex equation. Nevertheless, the above set of equations gets you fairly good results, otherwise consult graphical method (see Figure 14-8).

These equations show that turbulent noise is generally benign. Note, that turbulent noise increases to the fifth power of jet velocity at the valve orifice. This is proportional to 25 log (ΔP).

Noise Produced by Turbulence and Cavitation of Liquids

$$L_{a,liq.} = A + B + C \qquad \text{in dBA} \tag{14-26}$$

While turbulent noise is relatively simple to estimate, the complexity of noise produced by cavitation of water is extremely complex and equations provided by references 8 are based on a mixture of scientific laws and factors based on empirical data. No exact results can be expected due to the great number of variables such as the number of entrained particles in the water, the surface roughness of the body wall and so on. Rather than reproducing such many and complicated equations I thought it best to provide my readers with a simple graphical method, where,

Here A is a base level found in the graph below; B is a pressure adjustment factor, where $B = 10 \log(P_1/100)^{1.625}$, where P_1 = inlet pressure in psig. C = is a pipe wall correction factor, where $C = +4$ for schedule 20, 0 for schedule 40, -4 for schedule 80 and -8 for schedule 160 (in dB).

Here is an example: What is the max sound level of a 4″ globe valve at an inlet pressure of 1500 psig and 800 psig outlet pressure? The pipe wall is schedule 80. First we calculate the pressure drop which is 1500 – 800 = 700 psi. This makes the pressure ratio 700 / 1500 = 0.47. Looking at Figure 14-8, we find an A value of 90 (extrapolating between 3″ and 6″). Value B is $10 \log(1500/100)^{1.625}$. This is 19.1 dB. Finally C is -4 for schedule 80 from the above table. The expected total sound level therefore is: 90 + 19.1 – 4 = 105.1 dBA.

The value of X_{Fz} can be calculated using equation 14-19.

AVOIDANCE OF LEAKY VALVE STEM PACKINGS

A valve stem packing is one of the most important elements of a control valve, yet one that is not fully understood. One thinks, quite correctly, that a worn stem packing causes external leakage of fluid passing through the valve, but how many of us realize that a tight packing can make the controller go unstable? The combination of low friction and low stem leakage is what makes a good packing. Notice that I said low leakage, not zero leakage, since there is no such thing as zero leakage.

In former times when asbestos was the standard packing material, packing boxes had lubricators whereby the maintenance people could inject grease into the packing, which not only reduced friction but also provided a high viscosity

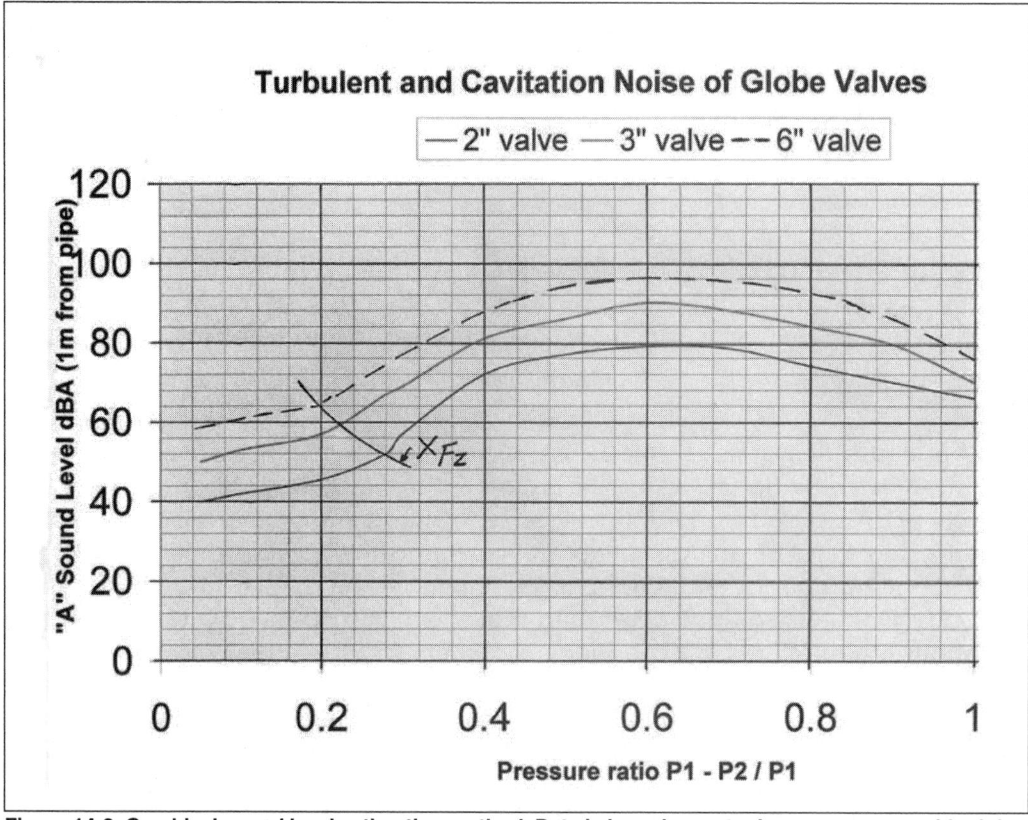

Figure 14-8. Graphical sound level estimating method. Data is based on actual measurements with globe valves (eccentric rotary plug valves are also applicable) at 70% of rated flow, schedule 40 pipe and 100 psig inlet pressure. The medium is water at room temperature.

interface between the packing bore and the stem exterior to improve seal ability. Any maintenance department had to stock quite a variety of packing greases with varying viscosities (depending on service temperature), especially composed to resist corrosive attacks by certain fluids. Teflon® changed all that and the all-Teflon V-ring—or Chevron packing, preferred by some, and the TFE-lubricated asbestos packing preferred by others—became the industry standard. Now we do have a variety of plastic materials such as Kalrez® available.

While the V-ring packing could not be repaired without taking the whole valve apart (which usually meant shutting down the process), it offered constant stem friction, especially when preloaded by a compression spring. The TFE-asbestos packing or its current successors, the TFE-graphite or TFE-Kevlar® packings, are made in the form of a split-ring and allow the emergency repair of a valve in the line by slipping one or two extra rings under the packing follower. The disadvantage of this type of packing was, and still is, that its tightness depends on the "feel" of the maintenance person, and many a carefully tuned process controller has gone unstable because of added stem friction (increased valve dead time) due to an over tightened of the stem packing.

Here then are some rules on how to avoid trouble with a valve stem packing:

- Make sure the packing is tight *before* placing the valve in operation. Some manufacturers ship the valve with a loose packing flange or nut.
- If hot fluids are involved, retighten packing after valve has fully warmed up when first placed in service.
- During installation, do not rotate the actuator (to change positioner location, for example) without loosening the packing before moving the actuator. We have to realize that the center of the stem and the center of the packing are not perfectly aligned. A loose packing will adapt to a change in center.
- Do not allow your control loop or actuator to become unstable for a long time. Remember, even one cycle per second cyclic motion can translate into 2 1/2 million cycles of packing wear in the time frame of only one month! The packing is completely worn at that time.
- Do not use packing material that has a high coefficient of thermal expansion when handling steam or hot fluids.
- Total effective packing depth should always exceed the length of valve travel. This will avoid creating a groove inside the packing bore in case there is some "nick" on the surface of the valve stem.
- Packing for rotary valves is superior in performance to those for reciprocating stem valves as long as the shaft of the rotary valve is well supported to absorb the perpendicular actuator thrust or the off-center strain caused by a misaligned stem-coupled rotary actuator.
- Misalignment between valve stem and actuator stem causes wear and friction. This can happen when an actuator yoke, made from steel or aluminum, bends following a drop during shipment. Cast iron yokes cannot do this - they break. Misalignment can also happen when a heavy actuator is mounted in a horizontal position without proper support.

While no packing is tight, remember, the permeation rate[9] through *solid* steel alone is already within the measurement sensitivity of mass spectrometers. So don't expect too much. Here are some rough examples of what you can expect in form of packing leakage:

A typical 5/16 in. stem diameter packing with solid TFE Chevron rings (preloaded) exhibited a stem leakage when subjected to 55 psi air at 70°F of about 0.4 scc/min after 100,000 full travel cycles, or about 550 scc/day, which is roughly 2×10^{-5} Kg/hr and is still below the 500 parts/million limit. The same valve exhibited a leakage of 1×10^{-5} Kg/hr of water under the same conditions. The above test data is based on laboratory measurements and exclude environmental effects such as corrosion or dirt, which will make things worse in real life.

In a stable control loop situation, 100,000 full travel cycles are a great many, and a valve packing may not need to be replaced for a number of years. In other cases, a tight maintenance schedule has to be instituted. On the other hand, you may want to consider that the packing wear equivalent to 100,000 x 1/2 in. travel cycles is 625 days of operation. However, an unstable valve having average stem excursions of ±030 in. at a rate of two cycles per minute accumulates 1.05 million cycles in a year! This demonstrates again the need for a stable loop and a stable actuator.

When all else fails, specify a packless valve such as a diaphragm valve (metal or elastomers), a pinch valve (elastomers only), a flexible sleeve valve (TPE, plastic, or elastomers), or a bellows seal valve.

BELLOWS SEALS

The use of bellows is one form of avoiding valve stem leakage and a way to meet EPA-recognized avoidance of hazardous leakage from valves[12] (the others are use of diaphragm and pinch valves).

Bellows seals work well if properly maintained and not overstressed. They should *always* be used with a backup packing and a telltale connection between the bellows seal and packing, leading to a pressure switch or pressure gage to inform or alarm the operator in case of bellows failure. Bellows pressurized from the outside typically give higher pressure ratings.

Finally, double walled bellows are available for added leakage protection. Note: removeability of the bellows requires a good gasket at the interface between the static portion of the bellows and the bonnet.

The following are causes of bellows failure.

OVER PRESSURIZATION

This is usually caused by subjecting the valve *with* bellows to hydrostatic line test pressure for which typical bellows are not designed (consult the manufacturer). Other sources of excess pressure are pressure waves such as "water hammer," which is caused by sudden valve opening, or thermal expansion of liquid trapped in the valve and pipe due to ambient temperature changes (a relief valve may be required).

FATIGUE

This can be caused by excess valve cycles. Note, most bellows are rated for a minimum of 100,000 full travel cycles, which means many years of normal throttling service (two *full* open-to-closure cycles per day will yield 100,000 cycles in 137 years!). However, if this valve is used in an on-off mode at, say, one cycle every five minutes, 100,000 cycles are reached in about one year.

However, fatigue problems can be more subtle. For example, having the bellows subjected to the pressure fluctuations of a positive displacement pump operating say at 100 cycles per minute can *partly* deflect the individual bellows convolutions at a rate of 1 million cycles per week and subsequently break the bellows (here a pulsation damper downstream of the pump is the answer). An unstable, cycling electronic control signal to the actuator can do the same damage by subjecting the stem and bellows to millions of minute cycles.

MECHANICAL DAMAGE

This is usually caused by improper installation or maintenance. The most common problem is twisting of the bellows when the valve stem is turned. A good valve design prevents the possibility of twisting the bellows during assembly or disassembly of the valve. Remember: Instructions are read only *after* things have gone wrong! Also, be careful when removing the actuator, you may inadvertently push or pull the valve stem (with the attached bellows) above the normal travel distance.

Pressurization of the *exterior* of the bellows increases resistance to working pressure by about 50%. However, this requires the retaining bonnet to be pressure and/or corrosion resistant. Adding balancing pressure to the *inside* of the bellows will increase the pressure rating. However, the fluid has to be inert and, in case of liquids, adequate relief must be provided to allow for venting of displaced volume during valve travel.

REFERENCES

1. Carucci, V. A., and Mueller, R. T., "Acoustically Induced Piping Vibration in High Capacity Pressure Reducing Systems," ASME PAPER 82-WA/PVP-8 (1982).
2. Baumann, H. D., "On The Prediction of Aerodynamically Created Sound Pressure Level of Control Valves," ASME PAPER W A/ FE-28, (1970).
3. IEC Standard 60534-8-3 (2000), "Control Valve Aerodynamic Noise Prediction Method," International Electrical Commission, Geneva, Switzerland.
4. Reed, C., "Optimizing Valve Jet Size and Spacing Reduces Valve Noise," *Control Engineering,* pp 63-64, Sept. (1976).
5. Singleton, E. W., "An investigation into control valve-generated turbulence and noise," *Measurement + Control,* Volume 37, Issue 6, July 2004.
6. Dobben, T., "Kavitation in Stellglieder," Frankfurt, Germany: Samson A.G.

7. Baumann, H. D., "A Practical Guide For The Cavitation Prevention of Throttling Butterfly Valves," ISA PAPER C.1.84-R777, Research Triangle Park, NC: ISA, Oct. (1984).

8. Baumann, H. D., Kiesbauer, J. "A method to estimate hydrodynamic noise produced in valves by submerged turbulent and cavitating water jets." *Noise Control Engineering Journal*, Volume 52 -2. March – April 2004. pp. 49-55.

9. Baumann, H. D., "Should a Control Valve Leak?," *Instrument And Control Systems*, August (1966).

10. Baumann, H.D., "Predicting Control Valve Noise at High Exit Velocities," *INTECH*, pp 56-59, February (1997).

11. Page, George W. Jr., "Predict Control Valve Noise," *Chemical Engineering*, pp 137-142, July (1997).

12. OSHA 1910.119-1992, "Process Safety Management for Highly Hazardous Chemicals," Washington, DC, Occupational Safety and Health Administration, 1992.

15

SEAT LEAKAGE AND SEAT MATERIALS

In order to save money and piping space, most users look for a control valve that can double as a shut-off valve. Alas, this is hard to accomplish.

First, in order to be "drip tight" (ANSI/FCI Class VI), there has to be a soft interface between the seat ring and plug, such as an elastomer or a plastic material. This then excludes high temperature applications.

Second, control valves due throttle all the time. This means that seating surfaces can be subjected to very high fluid velocities that can cause erosion. Quite often there are entrained particles within the controlled fluid which hasten the wear. This implies that whatever the manufacturer certifies as measured leakage when the valve left the plant may not be applicable when the valve is put in service.

Nevertheless, there are applications where a control valve can be used as a shut-off device, even if only in an emergency. A steam by-pass valve in an atomic power plant comes to mind. There are other non-critical loop applications where a small leakage can be tolerated and where the amount of leaked fluid has no material impact on the process and there is no safety hazard.

Another criterion is rangeability. You don't want the minimum flow rate that you want to control to be less than the leakage rate. This can be a consideration with small flow valves or swig-through butterfly valves. The type of seat material quite often determines the expected seat leakage rate. As mentioned before, a soft seating material will provide the lowest leakage rate, typically Class VI of ANSI B16.104 -1976 (FCI 70-2). Elastomers such as Buna N™ are used for moderate temperature (to 180°F), while Teflon™ is widely used when corrosive fluids are involved. Here the temperature limit is 450°F. Attention should be given if Teflon™ is used as seating material in certain rotary valves such as butterfly or ball valves. Here the problem is "sticktion" or "break-away torque" that is the difference between static and dynamic friction. It may take considerable actuator force to overcome this static friction leading to temporary valve instability.

The same may be said about some metal-seated valves such as high-performance butterfly valves, although the problem is less severe. Metal-to-metal seat valves are used in nearly all globe style control valves. For low pressure drops (up to 500 psi), Type 316 stainless steel is the preferred choice where the usual leakage rate for single-seated valves is Class IV and for double-seated valves Class II. Type 440 C is a hardened stainless steel often used for steam applications up to 800°F. A cobalt alloy such as Stellite™ is the preferred seat material for higher temperatures (up to 1500°F) and high pressures. Again the typical seat leakage (single seat) conforms to Class IV although, as with the Stainless steel seating, Class V can be met with special lapping techniques.

Table 15-1. Control Valve Seat Leakage Classifications per ANSI B16.104 (FC170-2)[a]

CLASS	Allow. Leakage	Test Medium	Test Pressure drop
I	No test required		
II	0.5% of rated capacity	Air or water at room temperature	45 – 60 psi to atmosphere
III	0.1% of rated capacity	Air or water at room temperature	45 – 60 psig to atmosphere
IV	0.01% of rated capacity	Air or water at room temperature	45-60 psig to atmosphere
V	0.0005 ml per minute of water per inch of seat diameter (D) at max. service pressure drop	4.7 ml per minute of air or nitrogen per inch of seat diameter (c)	50 psig to atmosphere (c)
VI	see tabulation	Air or nitrogen at 50-125°F	50 psig or oper. pressure[b]

a Revised per IEC Standard 60534-4-4 (2008)

b If lower than 50 psi

c. Alternate air or N2 test

Table 15-2. Class VI Leakage Rates

D) Seat Diameter in inches	Leakage Rate in ml per minute (cc/min)	Bubbles per minute
1 (25 mm)	0.15	1
1-1/2 (40)	0.30	2
2 (50)	0.45	3
2.5 (65)	0.60	4
3 (80)	0.90	6
4 (100)	1.70	11
6 (150)	4.00	27
8 (200)	6.75	45
10 (250)	11.1	–
12 (300)	16	–
14 (350)	21.6	–
16 (400)	28.4	–

The Class V leakage specification using gas as a test medium is a relatively new standard. The allowable quantities are derived from the water test volumes assuming laminar flow conditions. Here is an example: seat diameter 4 inch. Allowable leakage is 4.7 x 4 = 18.8 cc/min.

16

VALVES FOR SANITARY OR ASEPTIC SERVICE

Sanitary control valves have been around as long as dairy products have been processed automatically. These are highly polished body subassemblies that can be taken apart rapidly and are washed after each use.

Control valves used in the bioprocessing industry are another matter. While the former only needs to be "clean," the bioprocessing valve has to be germ free (aseptic)!

In contrast to sanitary food-processing valves (see Figure 16-1), bioprocessing valves are not disassembled but "cleaned-in-place" (CIP) and "sterilized-in-place" (SIP)[3]. Quite often, such valves are welded by orbital welding techniques to the pipe.

Modulating-type control valves, in contrast to automated on-off valves, are seeing increased use, especially in bioprocess systems that are upgraded from small batch systems to larger production systems. Applications range from pressure reduction of sterile steam to injection and flow control of high purity water, in quantities from 0.001 to 500 liters per minute. Design considerations for such valves need to meet CIP and SIP requirements[1] and be able to provide good flow characteristic, high rangeability, tight shutoff, be cavitation resistant, self-draining, and have low operating friction in order to meet the requirements for a stable control loop calling for minimum dead time by the controlling element.

Early valve types used in bioprocess equipment were originally designed for the food industry under the guidelines of the 3-A Council in order to meet "sanitary" requirements[2].

The angle valve shown in Figure 16-1 is a typical example. This valve will meet all requirements of a sanitary design. It is self-draining, can be polished to the required surface finish, usually has an R_a of 0.15 to 0.20 μM, and can have sufficient seal integrity at the interface between the upper housing flange and the bonnet guiding the stem. However, one difficult area to sterilize and to clean for aseptic service is the stem O-ring seal. Here a continuous cycling of the valve during the sterilization process is perhaps the only way to accomplish this task, calling for an auxiliary timing control system. The valve plug provides an acceptable flow characteristic. However, the "flow-to-close" fluid flow direction will encourage pressure recovery (low FL), i.e., higher throttling velocities for a given pressure drop and the tendency to cavitate with liquids. Also, metal plugs will not provide tight shutoff. A more modern type is shown in Figure 16-2 where the instability prone "flow-to-close" has been replaced by a stable "flow-to-open" design. Here the o-ring seal is located to be cleanable and to meet the requirements of the ASME bio-processing standard.[1] This valve can be used for both sanitary as well as for aseptic applications.

Another time-proven design is the diaphragm valve shown in Figure 16-3, which can also be used in non-sanitary applications. Here we have good shutoff capability and a construction that can meet the aseptic requirements of the system. A major drawback is the poor flow characteristic. Some improvements can be made using a two-step actuating device available from one manufacturer to overcome this handicap. Since the fluid pressure is acting on a relatively large diaphragm area, actuator forces are high, thus requiring high air pressure piston actuators (and expensive valve positioners), or large (in comparison to the valve size) spring-opposed diaphragm actuators.

Another common design employs a diaphragm serving as a combination housing closure and valve sealing member as shown in Figure 16-4.[3] Here the flow characteristic is essentially linear due to the employment of a circular orifice, instead of a weir, and a relatively short diaphragm travel. The valve housing is able to meet all aseptic material and surface finish requirements and is self-draining. The high actuator force requirement of diaphragm valves is overcome by the employment of a unique stainless steel toggle mechanism inside the valve bonnet, which, incidentally, is equipped with a secondary seal and telltale connection to guard against diaphragm rupture. Of additional interest is the special flow profile of the diaphragm. Notice the hollow indentation above the orifice. This leads to a curved inlet path for the fluid into the narrow throttling area along the rim of the orifice. After passing this "throttling gap," the fluid is suddenly separated from the diaphragm due to the latter's approximate 45° retracted slope. This abrupt flow pattern avoids fluid attachment and reduces pressure recovery of the fluid jet ($F_L \geq 0.85$), hence, assuring a low tendency to cavitate.

This valve is typically equipped with an EPDM diaphragm that is PTFE (Teflon®)-faced by chemical bonding. The rolling action of the diaphragm prevents stretching, which would separate the PTFE coating. Experience has shown that this material combination survives the rigors of up to 160°C (330°F) steam service even when the valve is used in continuous steam pressure reducing service. The mechanical amplifying toggle mechanism also reduces friction from the moving internal valve parts. As a result, the typical "dead band" of the valve is less than 2% of signal span, which allows the valve to be used without a separate valve positioner and the associated higher air consumption.

One negative aspect of this construction is the need to mount the actuator in the horizontal plane in order to achieve drainage, which calls for actuator support in larger sizes.

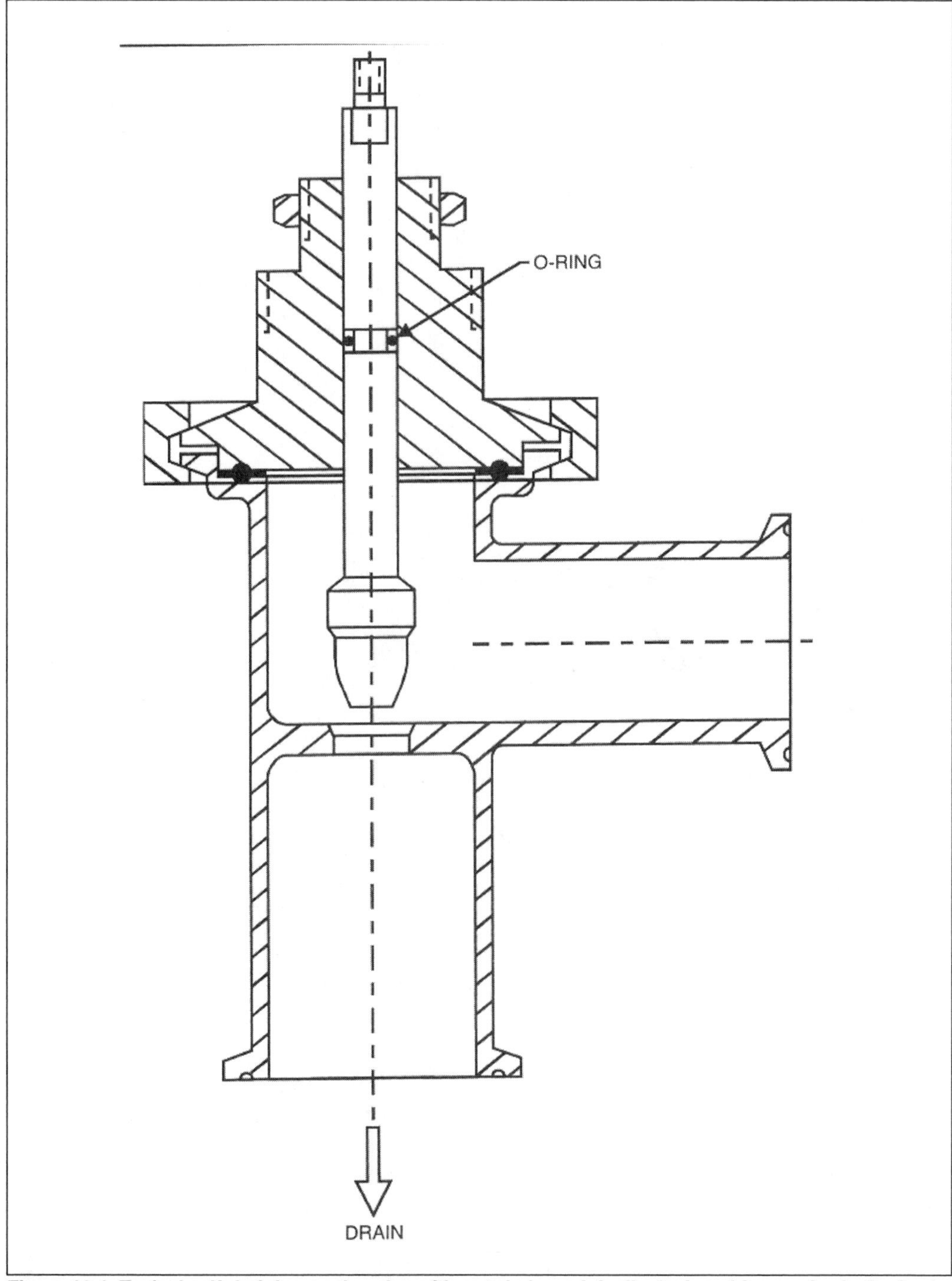

Figure 16-1. Typical self-draining angle valve with metal plug originally designed for service in the food industry under the guidelines of the 3-A Council.

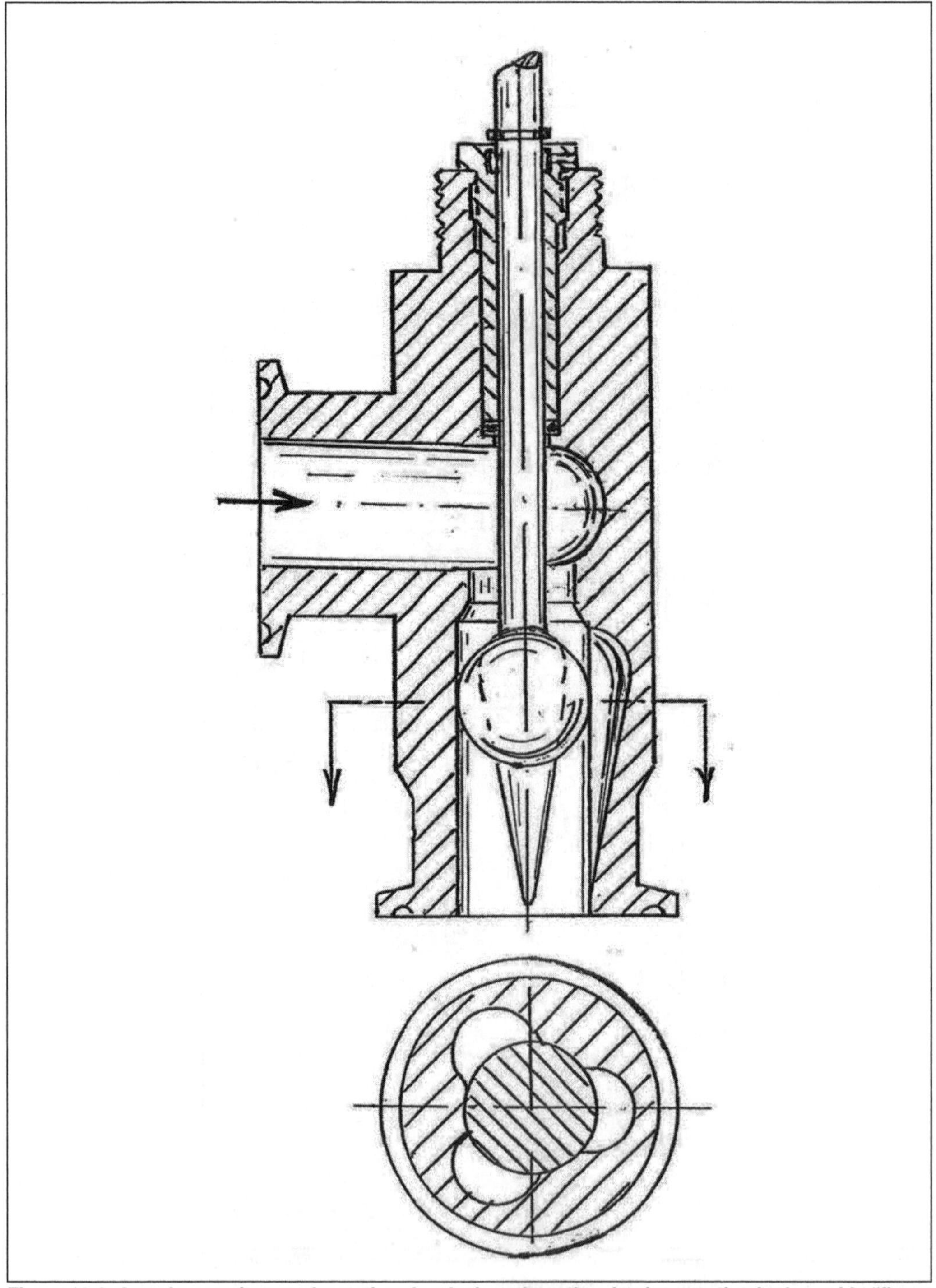

Figure 16-2. A modern sanitary and aseptic valve design where the plug is operating in the stable "flow-to-open" direction. Tapered grooves provide easy to polish and to clean flow passages. Number of grooves determine flow capacity. The flow characteristic is about "equal percentage". A clamped in seal (not shown) will provide Class VI shut-off. A cylindrical plug replaces the polished ball in larger valve sizes. Reference: US Patent 7,201,188.

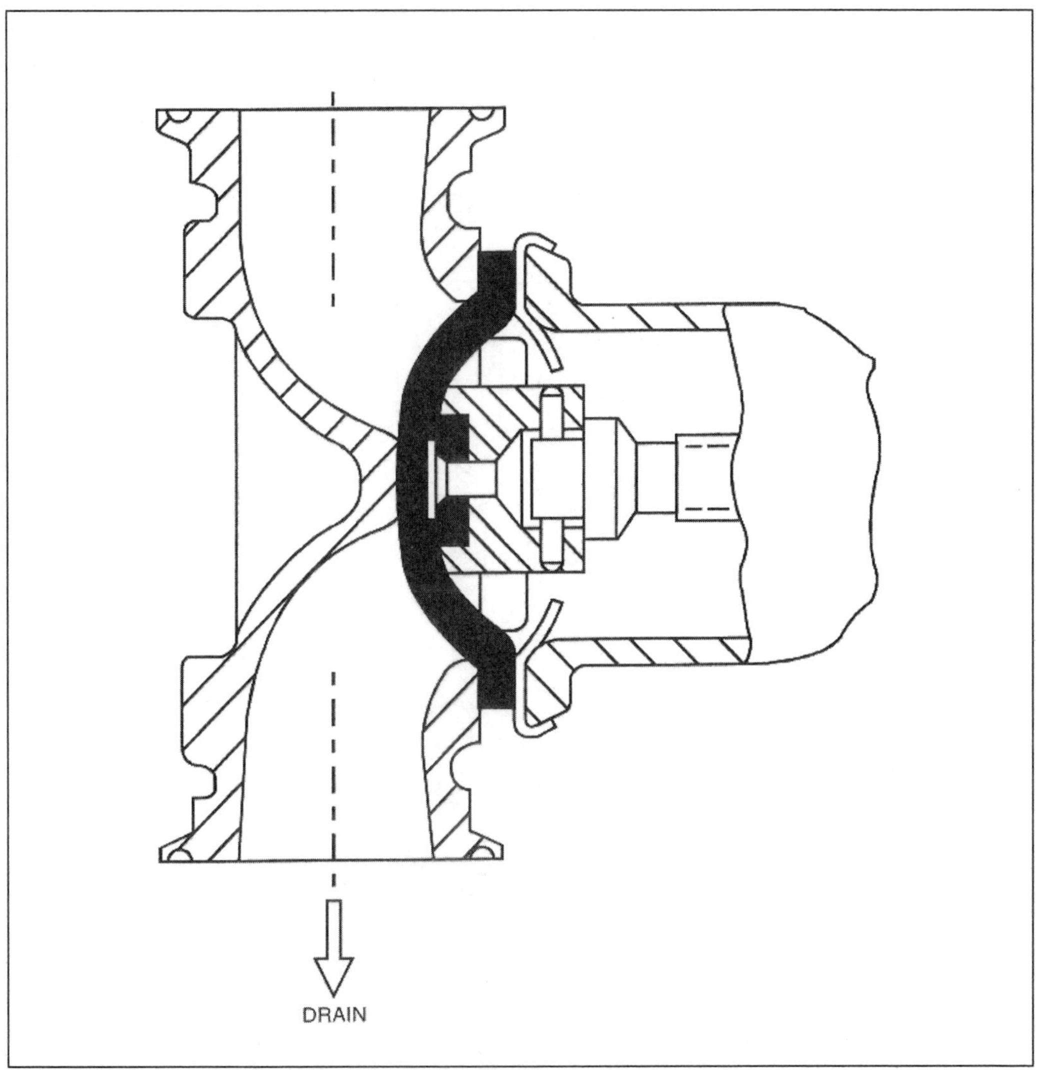

Figure 16-3. Diaphragm valve shown in a typical mounting position provides good shut-off, but poor flow characteristic.

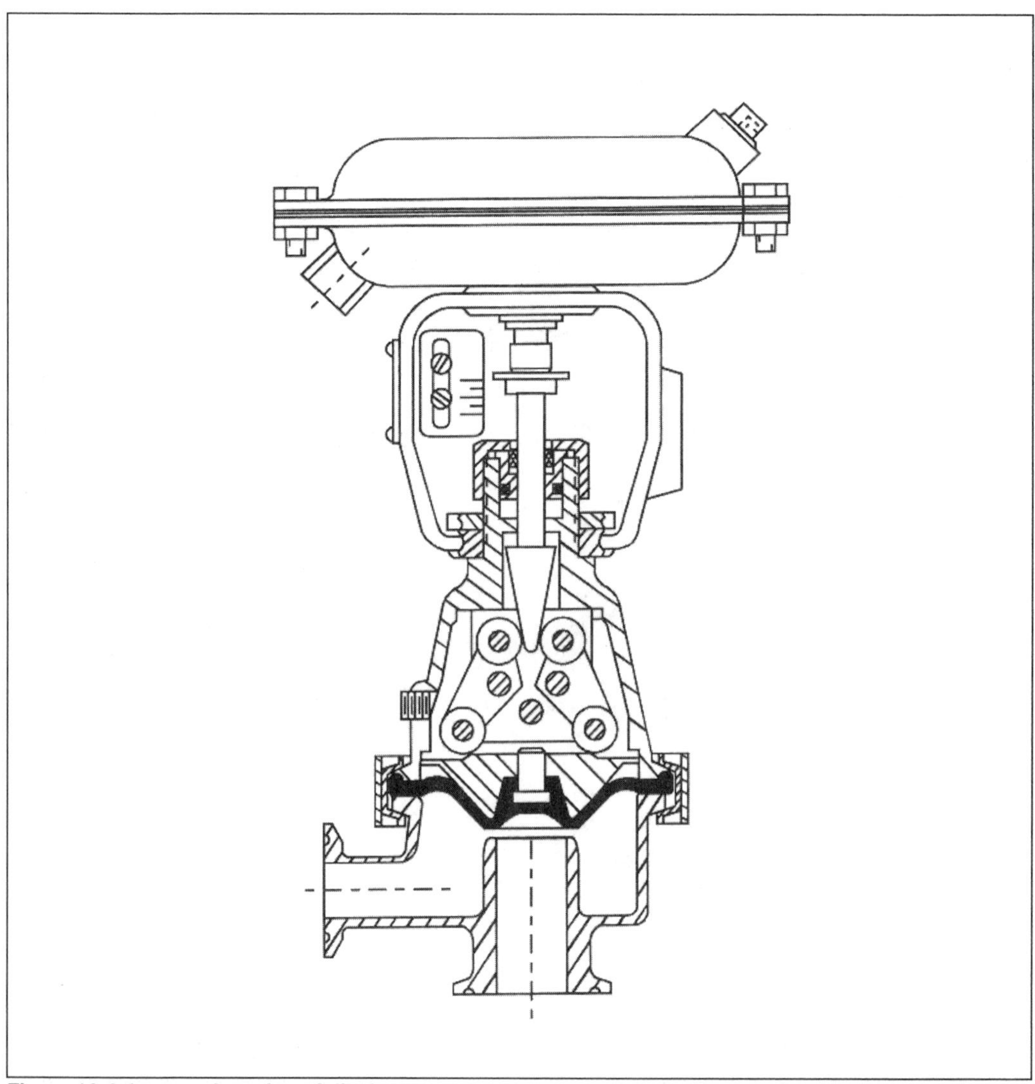

Figure 16-4. Improved version of diaphragm style valve designed especially for modulating control in aseptic service. Diaphragm is available with bonded PTFE over EPDM elastomer. (Courtesy of Baumann Inc.)

Table 15-1. Surface Finish Comparison[2]

Roughness Average R_a				
Grit Finish	μ inch	μ meter	RMS	ISO No.
150	30 - 35	0.8 - 0.9	45	N6
240	15 - 20	0.4 - 0.5	17	N5
320	8 - 12	0.2 - 0.3	9	N4

REFERENCES

1. ASME Standard BPE-1997, Bioprocessing Equipment. New York, NY: The American Society of Mechanical Engineers.
2. Ciancarelli, J. D., "Sanitary Control Valves for the Biotech Industry," *InTech*, pp. 18-21, May (1991).
3. Baumann, H. D., "Construction and Application of Modulating Control Valves for Bioprocessing Systems", Bioprocessing Engineering Symposium, ASME BED-Vol. 21, pp. 93-96 (1991).

17

FIFTEEN COMMANDMENTS: WHAT YOU SHALL NOT DO!

You shall not:

1. Install a steam valve at the bottom of a down-header.
2. Install a steam valve at the bottom of a down-header without providing a steam trap near the inlet side of a valve.
3. Locate a condensate return valve more than three feet away from a flash tank or condenser.
4. Buy a cage-trim valve for elevated temperature service without elastic support between cage and bonnet.
5. Buy an I/P transducer with a 3-15 psi output signal when the actuator spring of the valve is also 3-15 psi and there is no positioner.
6. Buy a flangeless, wafer-style control valve without a protective shroud where there is a fire hazard.
7. Start up a control valve without first checking the packing tightness.
8. Hydrostatically test a valve with bellows seal without first checking with the manufacturer.
9. Install a single-seated globe valve with spring-diaphragm actuator, flow-to-close, unless it has a small flow trim or it is used only for on-off service.
10. Use carbon steel valve body material for saturated water (condensate) service.
11. Install a valve in a pipe more than twice the diameter of the valve.
12. Let a controller go unstable for too long; the result may be a leaky stern packing.
13. Allow the valve outlet velocity exceed 250 ft/sec when you use a low-noise trim.
14. Install a horizontally placed heavy actuator without proper support.
15. Insulate a valve bonnet on a cryogenic valve.

18

ELECTRIC VERSUS PNEUMATIC ACTUATORS

Why not use electric actuators? This has been an illusive goal for the past 40 years. In 1965, when electric actuators started to be used, a major market research company predicted that 85% of all control valves would be electrically operated by 1975! Alas, this did not happen. The numbers are still less than 5% of the total. So much for forecasting market trends.

There are definite applications where electric actuators offer advantages over pneumatic actuators[1]:

- where precise positioning with no dead band is required (actuators with up to 0.02% of travel resolutions are available);
- where there are requirements in remote locations and no instrument air is available;
- to overcome severe dynamic forces acting on a valve stem; and
- where high stem forces are required (usually coupled with a low frequency response).
- for valves having high packing friction, or break-away torque.
- where the valve must fail in the "as is" position upon power failure.

The following major electric actuator types are available.

Electric Gear Drive Actuators. These actuators typically utilize AC motors over 1000 rpm where the motor revolutions are gradually reduced to less than 10 rpm by internal gear drives ending, in case of reciprocating drive mechanisms, in a rotating threaded nut that moves the spindle up and down. These actuators are reasonably inexpensive and can withstand wide temperature ranges but do not tolerate unstable control loops (see below). Note, the duty cycle may be as low as

25%. These devices have either linear or rotary power output (up to 5000 ft-lbs. torque).

Stepping Motor Actuators. These devices use DC stepping motors that can be positioned with up to 1 in 5000 steps by suitable amplifying circuits. The advantages are more precise positioning and the fact that the motor can be stalled without burning out. However, they are somewhat more expensive and are limited in thrust or torque due to the unavailability of large stepping motors.

Electric-Hydraulic Actuators. These devices use an intermediary power source (pumped oil) to drive a piston up and down. The oil is either pumped continuously or pumped on demand by a stepping-motor-driven pump. These devices are fairly rugged but consume more power, cost more, require auxiliary heaters to work in low ambient temperatures, and need more maintenance.

More advanced models use stepping motors to pump oils in discreet steps. This makes the device more rugged. Other advantages are that, like pneumatic actuators, they offer 100% duty cycle and their pistons can be fitted with a return spring, thus providing "fail-safe" action. Finally, they require less power consumption than continuously pumping models.

Modulating Solenoids. These actuators employ a spring-loaded, low power solenoid with a rather long armature. The valve stem is attached to an iron core which slides up and down the core of the wound armature following a variable voltage signal from a control circuit. These devices have relatively low power and therefore use pressure balanced trim designs. They are somewhat common in heating and air conditioning systems.

Most available actuators can be digitally controlled using either HART®, Foundation™ Fieldbus or Profibus architecture.

It becomes obvious, when comparing the implied complexity of such devices to a simple spring-loaded rubber diaphragm, why the pneumatic actuator still dominates. Other reasons why pneumatic actuators are preferred over electric actuators are: lower cost, more reliability, less bulk and weight, and higher frequency response (in smaller sizes). Another factor is the automatic availability of the fail-safe action provided by spring-diaphragm and some piston actuators (fail-open or fail-close of the valve).

One very important, but seldom recognized, reason is the excellent resistance of pneumatic actuators to minor upsets in the controller signal, which can be caused by a slight loop instability or even electrical noise. Even if such a ripple is passed through by a responsive I/P transducer, at an amplitude of perhaps ±0.01 psi of signal change and at a frequency of, say, 2 Hertz, these are completely absorbed by the relatively large air volume contained in a pneumatic actuator without causing so much as a tiny stem movement. In electrical terms, the pneumatic diaphragm case acts as a large capacitor!

In contrast, such signal ripples can be deadly in a gear-type electric actuator. Here, the motor will respond instantly (assuming a low dead band amplifier) causing a slight back and forth movement of the motor stem and the gears. This typically makes one tooth in a gear to constantly engage and to change load in both directions. After perhaps a million of such cycles, in a matter of weeks or months, that tooth is gone and with it the actuator.

REFERENCES

1. Béla G. Lipták, "Instrument Engineer's Handbook," Fourth Edition, Vol. 1: Process Measurement and Analysis, Research Triangle Park, NC: ISA/ CRC Press, p. 1824.
2. Kevin M. Hynes, "Electric Actuators Optimize Valve Response," *Power Engineering*, July 1988.

19

SAVING ENERGY

As I pointed out in Chapter 5, you should only use 5% of the total system pressure or 5 to 10 psi as allowed pressure drop across the control valve at the maximum flow rate. It will save your company a lot of electrical energy for pumps and compressors. This is contrary to the old rule of using about 50% of the dynamic head pressure in order to achieve a more linear installed flow characteristic (see Chapter 8).

Since the power converted in a control valve is similar to that used in a pump to create the pressure in the first place, it can be calculated as

$$P = 4.2 \text{ x } 10^{-4} \text{ x } \Delta P \text{ x } q \quad \text{in kW}$$

where:

q	=	flow in U.S. gpm of liquid
ΔP	=	pressure drop in psi

For example, in a valve application for 120 gpm oil at G_f *of* 0.8, the extra pump horsepower needed (a bigger motor is needed) if 50 psi pressure drop is assigned to the valve:

$$P = 4.2 \; x \; 10^{-4} \text{ x } 50 \text{ x } 120 = 2.5 \text{ kW}$$

To reduce the pressure, you need a valve size of only 1 1/2 inches. The C_v required is 15.2.

By reducing the valve pressure drop at max. flow to only 5 psi and choosing a different pump, the energy required for the valve is now only 0.25 KW, a savings of 2.5 – 0.25 = 2.25 kW per hour. This amounts to 20,000 kW per year! True, the valve size now has to go up to 2 inches to accommodate the higher C_v (48) based on only 5 psi pressure drop; however, the cost savings of about $2,000 per year more than pays for the cost difference between the two valve sizes. Besides, the line size for 120 gpm flow is 2 inches anyhow (see Table A-1), so you also save the cost of pipe reducers.

Another benefit besides energy (cost) savings is less valve repair, since there is no trim vibration that could be caused by higher pressure drop and no cavitation to wear out the trim.

In an article, Hans O. Engel[1] pointed out that the three largest chemical producers in Germany use about 120,000 pumps, with an average power consumption of 3 kW, cooperating with about 200,000 control valves. If all valves are sized more rationally (i.e., as proposed herein), then about half the power, or 180 MW of electric energy, could be saved. This, in turn, translates to a yearly savings of about 1.5 million MWh. It is fair to assume that the *total* U.S. chemical industry is about three times the size of the above German representation. This means a possible savings of 4 to 5 million MWh per year in the U.S.!

It is enlightening to realize how much energy a control valve converts when it throttles, i.e., when it does its assigned job of pressure reducing. The following equations provided by E. W. Singleton[2] allow you to do this.

For Liquids:

Non-choked

$$K_W = \frac{C_v}{2300}\sqrt{\frac{\Delta P^3}{G_f}};$$

Choked

$$K_W = \frac{C_v F_L \Delta P \sqrt{(P_1 - P_v)F_L^2}}{2300\sqrt{G_f}}$$

For Gases at Choked Flow:

$$K_w = 0.014\ C_v\ F_L{}^3\ P_1 \text{ Kilowatts}$$

where: All pressures are in psia

C_v = flow coefficient at desired flow conditions

F_L = pressure recovery factor at required travel

G_f = specific gravity at flowing conditions (water = 1.0)

The above should not be taken as an endorsement for speed-controlled pumps. These have their own problems, see Chapter 2.

REFERENCES

1. Engel, H. O., "Energieökonomie bei der Anwendung von Regelventilen," *RTP,* (Germany), No. 4:188-194, (1991).

2. Singleton, E. W., "Development of a High Performance Choke Valve with Reference to Sizing for Multiphase Flow," *Measurement and Control,* Vol. 24, No. 9, (1991).

20

THE BUS SYSTEM TO THE RESCUE, OR WHAT THE FUTURE MAY BRING

Will the conventional (throttling-type) control valve cease to exist? Not very likely. Remember, the automobile did not completely replace the bicycle. Obviously, it goes against our grain to "waste" energy in a throttling valve. However, despite attempts over the centuries to invent "perpetual motion," i.e., a device that works without "using" energy (converting energy to a higher entropy level), we have found none. The same is true for the control of fluids in a pipe. Basic laws of physics still apply.

Attempts have been made to obtain control more elegantly. For example, by applying electric current to a fluid,[1] you can reduce the flowing quantity by changing the turbulent or laminar flow pattern (see Figure 20-1). This works like a magnetic flowmeter in reverse; however, it is not a very practical approach. For the control of small flows, there are now micro-electronic valves available. One such design* relies on the interaction of a suspended mass with a valve body to open and close. In the closed position, the suspended mass is seated directly over an outlet port located in the valve body. Displacement of this mass opens the valve.

The displacement is achieved by changing the electric voltage supplied to a diaphragm, which is made of a bimetallic-silicon composite. The diaphragm surrounds the suspended mass to provide suspension. It is made up of resistors diffused into a thin layer of silicon and covered over by a layer of aluminum. When an electric voltage is supplied to the metallic materials, it effects a change in the temperature of the diaphragm and a corresponding displacement of the suspended mass.

The gap between the valve seat and its mating surface can be adjusted in proportion to electrical input, which allows proportional control of flow, as well as of the usual fully open or fully closed functions of a valve. The valve can control

* Manufacturer: IC Sensors

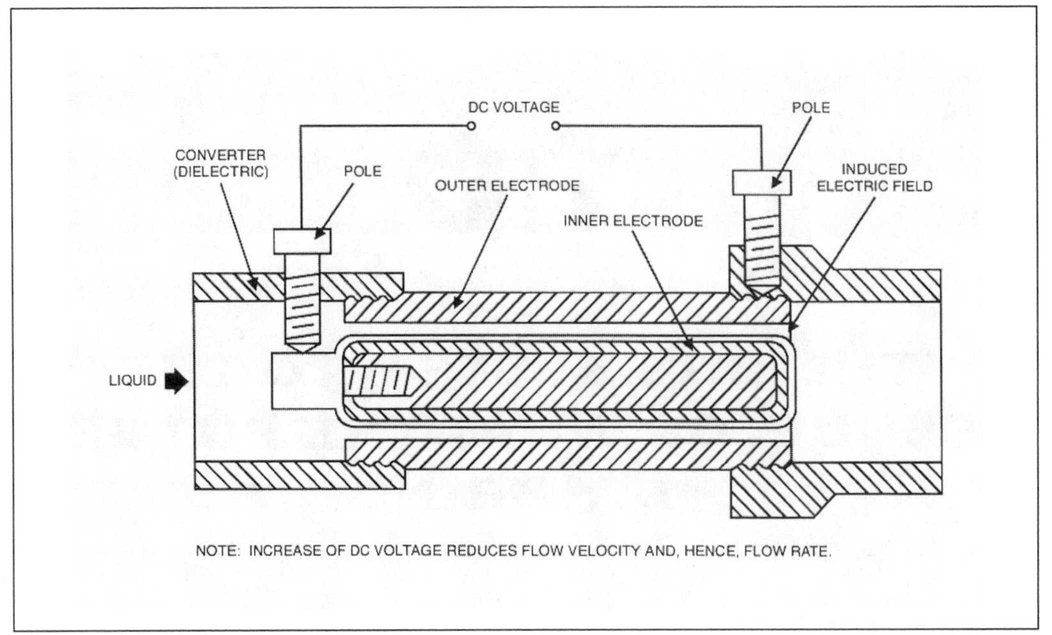

Figure 20-1. Solid-state control valve.

flow rates in the range of 0 to 150 cc/min at input pressures between 1 and 50 psig.

Speed-controlled pumps have been tried with limited success. So what is the future?

Well, electronics will still bring us important changes as far as actuating software and hardware is concerned. The biggest impact has been the use of fieldbus standards which now tend to replace the current analog 4-20 mA signal by a digital communication signal between the controller or computer and the valve positioner. The benefits is increased functionality (see Chapter 9).

Some positioners already have a PID function chip and do all the control based on the sensor input via the fieldbus. This reduces hardwiring costs (a single pair of twisted wire connects everything – see Figure 20-2), and offers simplified data transfer with less noise and increased accuracy. At the same time the fieldbus can also transmit important data back to the computer regarding the valve position, for example. Other sensors that could be embedded in a positioner could inform about seat leakage, stem packing friction, actuator friction, air leakage in the actuator, and so on. However, I pity the poor operator who must be absolutely overwhelmed by monitoring perhaps ten continuous data input channels per valve in a typical process plant having 500 individual loops! So, selective use of this option is certainly in order. Being informed of a higher than usual valve dead band could explain sudden loop instability and may trigger preventive maintenance, for example. Being aware of the stem package leakage could avoid dangerous spills.

More serious problems exist with the amount of electric current a fieldbus can handle without losing its "intrinsically safe" label. Of course, a low current means fewer valve positioners on the line. The solution is to multiplex (time share) so

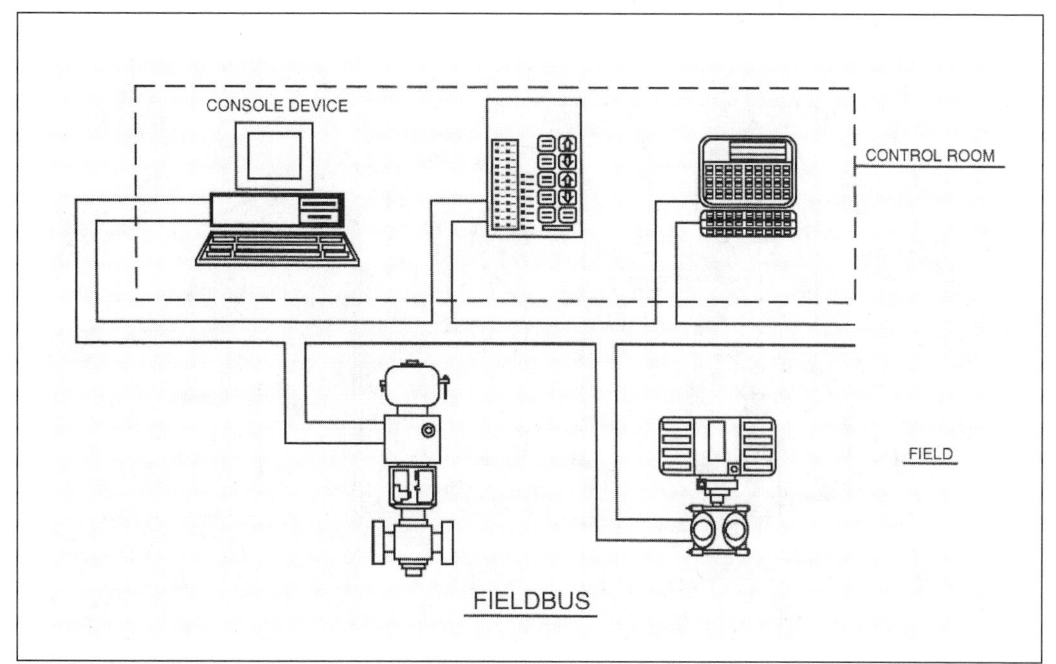

Figure 20-2. Schematic diagram of component connected in a fieldbus system.

that only one positioner in the loop receives a signal change at any given time. This will require a memory device or pneumatic lock-up device to freeze the valve positioner until the next "switch on," which in turn might slow the valve travel speed down to a level that may be unsuitable for flow or pressure control. This is one example of a number of problems that must be addressed prior to the wide acceptance of the fieldbus system. The likely solution would be to have separate cables to supply electric power to local actuating devices. This, of course, negates some of the advantages of having a fieldbus system, namely, savings in cable lines to individual components. Nevertheless, there is a basic goal that can be reached, i.e., tie in all local loop components such as sensors, controllers, and valves into an independent local working system (distributed control system) to be connected only for purposes of supervisory control and data logging to a central computer network. This will indeed result in major cost savings not only in cabling but also in the amount of data being transmitted.

"Smart valves," on the other hand, will have a future. The trouble is, what is a smart valve? My preferred definition is what I call "the ultimate in distributed control." Let me explain: First you have to realize that every control valve controls the rate of flow passing through it. Nothing else! The difference between, say, a level control valve and a pressure control valve is determined by the fact that the first one employs a level transmitter instead of a pressure transmitter. If the identical valve handling the same flow is in a loop where the feedback comes from a temperature transmitter, then we have a temperature control valve.

So, once we realize that each valve controls only the rate of flow, then we can go a step further and create a "closed loop" flow control system with the valve as its centerpiece. Referring to Figure 20-3, a flowmeter, preferably a vortex meter

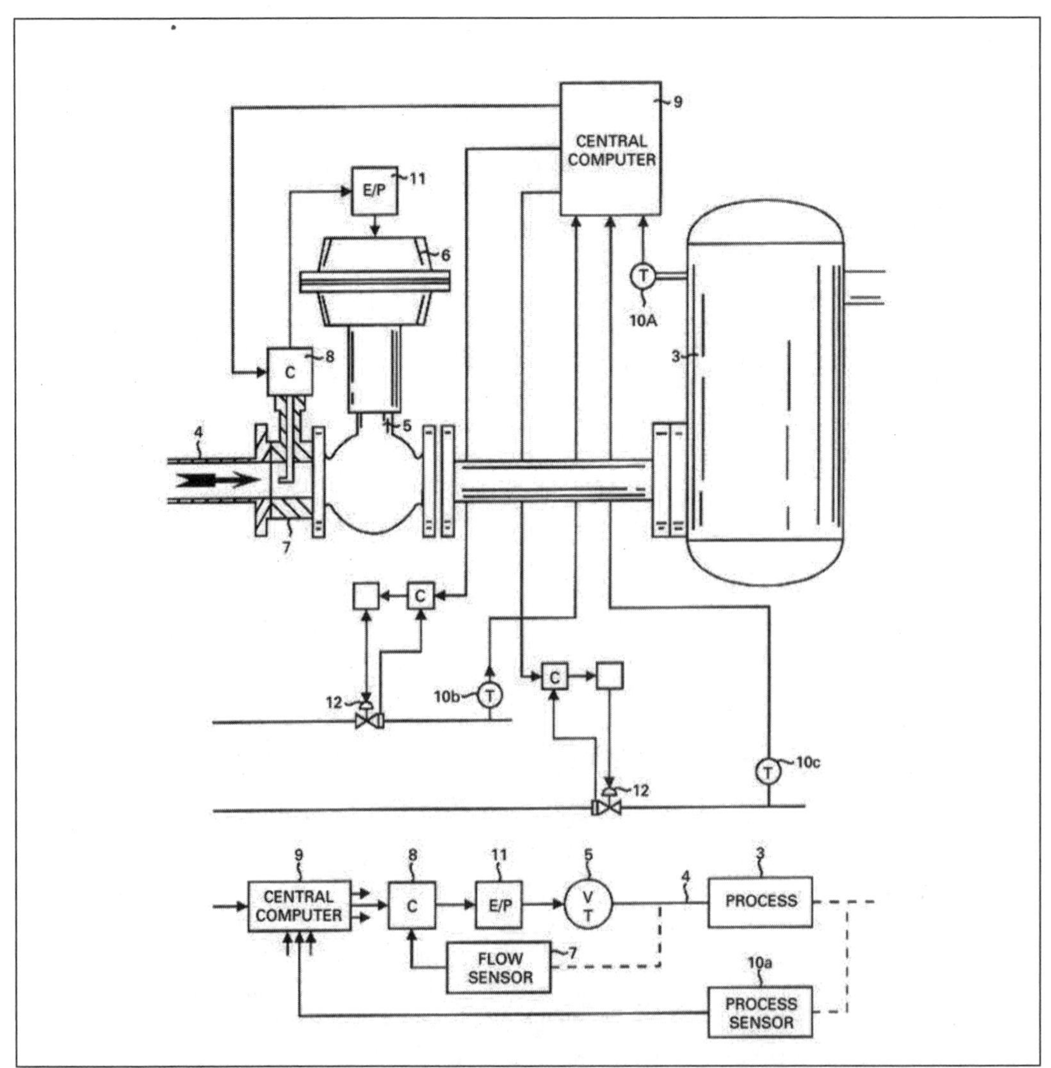

Figure 20-3. Patent drawing (by Author) of distributed control system where the control valve, in combination with a vortex-type flowmeter, fulfills primary flow rate controlling function in a temperature control loop.

(for range ability and a wide range of applications, including steam), is attached to the valve inlet (point of lowest turbulence). A flow controller is either attached to the flowmeter or is made part of the I/P positioner, while the set point for the flow controller is given by the main process controller or computer (shown as part of a temperature control system). The computer sets the flow rate to achieve the ultimate controlled temperature. As long as there is no disturbance in the system, everything will be fine. However, let's assume the heating medium passing the valve is steam, and, because of demand elsewhere, the inlet pressure drops by 10%. This can cause a sudden reduction in the steam flow rate by 10%, assuming choked flow. In a conventional temperature system it may take several minutes for the controlled temperature to begin to drop and for the computer to react and open the valve further. This leads to an upset in the controlled temperature.

With the smart valve, on the other hand, the flowmeter will sense the change in flow immediately, will make the valve increase the travel, and will restore the flow *before* the usually slow thermodynamic effects make themselves felt downstream. Thus, the temperature stays at the set point, and the computer probably never even detects the upset in steam pressure. This is, of course, a form of cascade control, except good portions of the loop are now in a tidy, compact package. Despite the added cost (and perhaps added wiring), a case can be made to use such a smart valve in critical loops where the valve can absorb process disturbances before they can affect the ultimate controlled variable.

Another favorite of mine, and of a few others,[3] is to resurrect the good old pressure regulator with its excellent frequency response and high flow range. Using a temperature control system similar to that described above, the regulator shown in Figure 20-4 can have a set pressure determined by the air signal from an I/P transducer signal coming from a temperature controller (see loop diagram). This regulator controls the pressure of a heat exchange fluid such as oil or steam ahead of an orifice (FO) or an adjustable restrictor. Sized correctly, the combination orifice area and pressure ahead of the orifice will effectively control the rate of flow (and, therefore, Btu) to the heat exchanger. Any demand for increase in temperature is met by the controller increasing the signal to the I/P transducer, which, in turn, increases the air load below the actuator diaphragm and causes the valve to open, thus increasing the pressure drop across the orifice, thereby, increasing the flow of heating fluid. Any local disturbance, such as a sudden decrease in supply pressure, is now sensed by the valve diaphragm, which leads to a nearly instantaneous correction without the disturbance being noticed by the temperature transmitter (TE) at the heat exchanger. Again we have a "smart valve" with nearly the same performance characteristics as those in Figure 19-3, but at a fraction of the cost.

THE COMING WIRELESS ETHERNET

The ISA-100 wireless standard currently being developed and the already published wireless HART® standard will make it easier for the user to convert serial networks to Ethernet communication. This will offer faster polling cycles and, of course, will eliminate a lot of wiring.

Problems to be solved are security, possible interference and noise. Wireless communication devices also need power, batteries, for example. Distance is also a problem (20 miles max.).

It is for these and other reasons that the initial use of wireless systems will be restricted to data transmission, and "open-loop" or "soft" close-loop control systems.

---

* For example, the temperature measurement lag of a 1/2 in. bulb thermal well sensing a heavy liquid at low velocities can vary between 60 and 180 seconds!

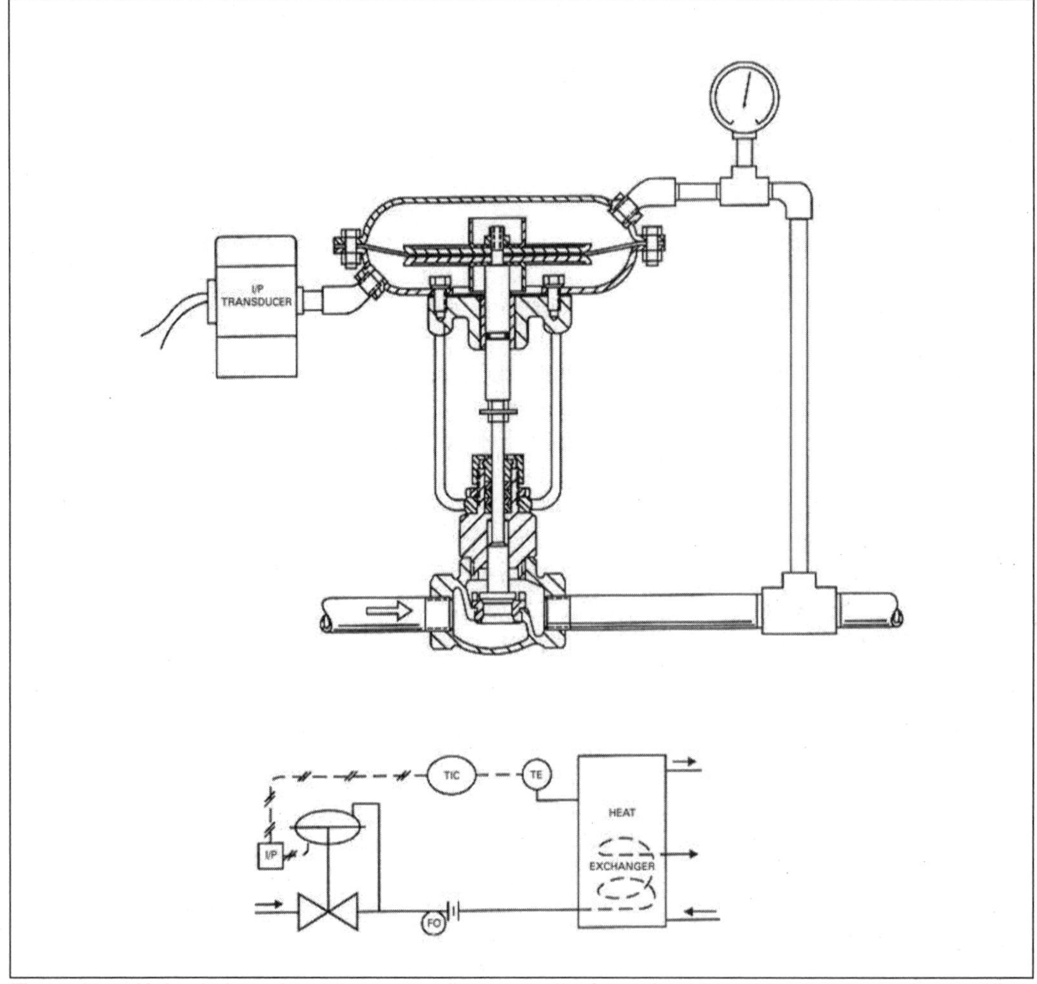

Figure 20-4. Air-loaded regulator serves as a flow controller in conjunction with a fixed or variable orifice (FO) as part of a temperature control loop .

THINKING "GREEN" AND SAVING ENERGY

There currently is a worldwide effort underway to save energy. This effort can be supported by users and manufacturers of control valves alike. Besides assigning lower pressure drops across valves as outlined in Chapter 18, emphasis could be placed in the design and selection of valve types. Most current designs are quite heavy and much metal* could be saved in cage valves, for example. Here the weight of the heavy bonnet flange could be cut in half with proper cage retention within the valve housing (see Figure 3-3). Many globe valves, too, can be replaced by rotary valve styles (not on-off types) that can perform the same job at less than half the weight. In addition, rotary control valves offer about 20% more flow

---

* According to an article in *wikipedia,* it takes about 2.5 KW to melt a pound of steel.

capacity (C_v) over globe valves, again offering savings in energy that is used to run pumps and compressors.

Such "green" thinking would not only drastically reduce the use of energy consuming metals but would also reduce transportation (freight) expenses. Finally, such savings will also greatly improve the profit margins of both manufacturer and user.

REFERENCES

1. Baumann, H. D., "Important Trends in Control Valves and Actuators," *Instrument & Control Systems,* pp. 21-25, Nov. (1975).
2. Cating, C. E., "Control Valve Industry Update-1990," *Control,* pp. 34-38, March (1990).
3. Driskell, L., "Save Money By Combining Regulators, Distributed Control," *Power,* April (1985).

BIBLIOGRAPHY

Beranek, Ver, *Noise and Vibration Control Engineering,* NY: John Wiley & Sons (1991).

Driskell, Les, *Control Valve Selection and Sizing,* Research Triangle Park, NC: ISA (1983).

Lipták, Béla *A., Instrument Engineers Handbook,* Fourth Edition., Taylor & Francis, NY. (2006).

Lyons, Jerry L., *Lyons' Valve Designer's Handbook,* NY: Van Nostrand Reinhold Company.

Control Valve Handbook, Marshalltown, IA: Fisher Controls International, Inc.

ISA Handbook "Control Valves", Guy Borden - Editor., Research Triangle Park, NC: ISA (1998).

"Flow of Fluids Through Valves, Fittings, and Pipes," Technical Paper No. 410, Chicago, IL: Crane Company.

Masoneilan Handbook for Control Valve Sizing, Avon, MA: Masoneilan- Dresser Co.

Standard and Recommended Practices for Instrumentation and Control, 10th Edit., Research Triangle Park, NC: ISA (1989).

Valves, Piping and Pipe Line Handbook, 2nd Edit., Surrey, United Kingdom: The Trade and Technical Press (1986).

APPENDIX A: REFERENCES—TABLES AND FIGURES

Table A-1. Flow Rates in U.S. GPM Based on Schedule 40 Pipe and 10 Feet/Second Fluid Velocity

Pipe diam.	1"	13/4"	2"	3"	4"	6"	8"	10"	12"
GPM	27	62	100	210	380	900	1600	2300	3500

Table A-2. Flow Rates in Pounds Per Hour of Saturated Steam at a Fluid Velocity of Approximately 150 Feet/Second in a Schedule 40 Pipe

	Pipe Diameters (Flow in lbs/hr)								
Pressure, psig	1"	1½"	2"	3"	4"	6"	8"	10"	12"
15	200	500	800	1800	3200	5400	12800	20000	24600
30	300	710	1200	2400	4800	9800	19200	30000	36000
60	500	1200	2000	4500	8000	18000	32000	50000	60000
100	750	1800	3000	6800	12000	27200	48000	75000	90000
150	1100	2500	4400	9900	18000	40000	72000	111000	132000
200	1500	3300	6000	13500	24000	54000	96000	150000	180000
300	2200	4700	8800	20000	32000	80000	128000	220000	265000
400	2500	6100	10000	22500	40000	90000	160000	250000	300000
500	3300	7800	13000	30000	52000	120000	210000	330000	400000

Table A-3. Common Materials for Valve Bodies

Metals ASTM	Designations
BRONZES AND BRASSES	
High Tensile Steam Bronze	B 61
Steam Bronze	B 62
IRONS	
Cast Iron	A 126 Class A
Cast Iron	A 126 Class B
Ductile Cast Iron	A 395
Malleable Cast Iron	A 47 Grade 32510
Ni-resist Gray Cast Iron	A 436 Type 2
STEELS	
Carbon Steel, Forged	A 105
Carbon Steel, Cast	A 216 Grade WCB
0.15% Moly. Steel, Cast	A 217 Grade WCl
Cr. Moly. Steel, Cast	A 217 Grade WC6
Cr. Moly. Steel, Cast	A 217 Grade WC9
4.6% Cr. Moly. Steel	A 217 Grade C5
8-10% Cr. Moly. Steel	A 217 Grade C12
Carbon Steel, Cast	A 352 Grade LCB
Carbon Moly. Steel, Cast	A 352 Grade LCl
3.5% Nickel Steel, Cast	A 352 Grade LC3
STAINLESS STEELS	
18 Cr. 8 Ni., Cast	A 351 Grade CF8 (Type 304)
Type 304 Bar	A 182 Type F304
Type 303 Bar	A 182 Type F303
18 Cr. 8 Ni. Mo., Cast	A 351 Grade CF8M (Type 316)
Type 316 Bar	A 182 Type F316
18 Cr. 8 Ni. Cb., Cast	A 351 Grade CF8C (Type 347)
18 Cr. 8 Ni. Cb., Bar	A 182 Type F347
18 Cr. 8 Ni. Mo., Forging	A 182 Type F316L
16 Cr. 12 Ni. 2 Mo., Bar	A 479 Type 316L
NICKEL ALLOYS	
Nickel Cast	A 296 CZ-l 00
Nickel Wrought	B 160
Monel Cast	A 296 M-35W
Monel Wrought	B 164 Class A
Hastelloy "B" Cast	A 296 N-12M
Hastelloy "C" Cast	A 296m CW-12M
ALUMINUM	
No. 356 T6 Cast	B 26 Grade SG70A T6
TITANIUM	
Alloy 20, Cast	A 296 Grade CN-7M
Cast	B 367
Wrought	B 381

Table A-4. Common Trim Materials

METALS	ASTM DESIGNATIONS
Brass Rod	B 16
Naval Brass	B 21 Alloy A
Bronze Rod	B 140 Alloy B
Phosphor Bronze	B 134 Alloy B-2
Type 316 St. Steel Rod	A 479 Type 316
Type 316L St. Steel Rod	A 479 Type 316L
Type 316 St. Steel, Cast	A 351, Grade CF8M
Type 316 St. Steel, Forged	A 182, F316
Type 304 St. Steel Rod	A 182, F304
Type 347 St. Steel Rod	A 182, F347
Type 17-4 St. Steel Rod	A 564
InconelRod	B 166
Monel Cast	A 296 Grade M-35

Table A-5. Common Bolting Material

	ASTM DESIGNATIONS	
Metals	Bolts	Nuts
Type 304 St. Steel	A 193 Grade B8, Cl. 1	A 194 Grade 8
Type 347 St. Steel	A 320 Grade B8	A 194 Grade 8
Type 316 St. Steel	A 320 Grade B8	A 194 Grade 8
Carbon Steel	A 193 Grade B7	A 194 Grade 2H
Carbon Moly. Steel	A 193 Grade B7	A 194 Grade 2H
5% Cr. 1/2% Mo, Alloy	A 193 Grade B16	A 194 Grade 4

Table A-6. Saturated Steam Pressure and Temperature

Vapor Pressure				
Absolute, Psia	Vacuum, In. Hg.	Temperature Degrees F.	Steam Density lbs./Cu. Ft.	Water Specific Gravity
0.20	29.51	53	.00065	1.00
0.30	29.31	64	.00096	1.00
0.40	29.11	72	.00126	1.00
0.50	28.90	79	.00156	1.00
0.60	28.70	85	.00185	1.00
0.80	28.29	94	.00243	1.00
1.0	27.88	101	.00300	.99
1.2	27.48	107	.00356	.99
1.4	27.07	113	.00412	.99
1.6	26.66	117	.00467	.99
1.8	26.26	122	.00521	.99
2.0	5.85	126	.00576	.99

Table A-6. Saturated Steam Pressure and Temperature

2.4	25.03	132	.00683	.99
3.0	23.81	141	.00842	.98
4.0	21.78	152	.0110	.98
5.0	19.74	162	.0136	.98
6.0	17.70	170	.0161	.98
7.0	15.67	176	.0186	.97
8.0	13.63	182	.0211	.97
9.0	11.60	188	.0236	.97
10.0	9.56	193	.0260	.97
12.0	5.49	202	.0309	.96
14.0	1.42	210	.0357	.96
14.7	0.0	212	.0373	.96
20	5.3	227	.0498	.95
22	7.3	233	.0544	.95
24	9.3	237	.0590	.95
26	11.3	242	.0636	.95
28	13.3	246	.0682	.94
30	15.3	250	.0727	.94
35	20.3	259	.0840	.94
40	25.3	267	.095	.94
45	30.3	274	.11	.93
50	35.3	281	.12	.93
55	40.3	287	.13	.93
60	45.3	293	.14	.92
65	50.3	298	.15	.92
70	55.3	303	.16	.92
75	60.3	308	.17	.92
80	65.3	312	.18	.91
85	70.3	316	.19	.91
90	75.3	320	.20	.91
95	80.3	324	.22	.91
100	85.3	328	.23	.90
105	90.3	331	.24	.90
110	95.3	335	.25	.90
115	100.3	338	.26	.90
120	105.3	341	.27	.90
125	110.3	344	.28	.90
130	115.3	347	.29	.89
135	120.3	350	.30	.89
140	125.3	353	.31	.89
145	130.3	356	.32	.89
150	135.3	358	.33	.89
160	145.3	364	.35	.88
170	155.3	368	.37	.88

Table A-6. Saturated Steam Pressure and Temperature

180	165.3	373	.40	.88
190	175.3	378	.42	.88
200	185.3	382.44	.87	
225	210.3	392	.49	.87
250	235.3	401	.54	.86
275	260.3	409.60	.85	
300	285.3	417	.65	.85
400	385.3	445	.86	.83
500	485.3	467	1.08	.81
600	585.3	486	1.30	.80
700	685.3	503	1.53	.78
800	785.3	518	1.76	.77
900	885.3	532	2.00	.76
1000	985.3	545	2.24	.74
1250	1235.3	572	2.90	.71
1500	1485.3	596	3.62	.68
2000	1985.3	636	5.32	.62
2500	2485.3	668	7.65	.56
3000	2985.3	695	11.7	.46

Table A-7. Temperature Conversion Table

°K	°C	°F	°R
0	-273	-459.4	0
33	-240	-400	60
61	-212	-350	110
89	-184	-300	160
116	-157	-250	210
144	-129	-200	260
172	-101	-150	310
200	-73	-100	360
227.4	-45.6	-50	410
230.2	-42.8	-45	415
233	-40.0	-40	420
235.8	-37.2	-35	425
238.6	-34.4	-30	430
241.3	-31.7	-25	435
244.1	-28.9	-20	440
246.9	-26.1	-15	445
249.8	-23.2	-10	450
252.4	-20.6	-5	455
255.2	-17.8	0	460
258	-15.0	5	465
260.8	-12.2	10	470

Table A-7. Temperature Conversion Table

263.6	-9.4	15	475
266.3	-6.7	20	480
269.1	-3.9	25	485
271.9	-1.1	30	490
273	0	32	492
274.7	1.7	35	495
273.4	4.4	40	500
280.2	7.2	45	505
283	10.0	50	510
285.8	12.8	55	515
288.6	15.6	60	520
291.3	18.3	65	525
294.1	21.1	70	530
296.9	23.9	75	535
299.7	26.7	80	540
302.4	29.4	85	545
305.2	32.2	90	550
308	35.0	95	555
310.8	37.8	100	560
313.6	40.6	105	565
316.3	43.3	110	570
319.1	46.1	115	575
322.9	48.9	120	580
327.4	54.4	130	590
333	60.0	140	600
338.6	65.6	150	610
344.1	71.1	160	620
349.7	76.7	170	630
355.2	82.2	180	640
360.8	87.8	190	650
366.3	93.3	200	660
377.4	104.4	220	680
394	121	250	710
422	149	300	760
450	177	350	810
477	204	400	860
505	232	450	910
533	260	500	960
561	288	550	1010
589	316	600	1060
616	343	650	1110
644	371	700	1160
672	399	750	1210
700	427	800	1260

Table A-7. Temperature Conversion Table

727	454	850	1310
755	482	900	1360
811	538	1000	1460
866	593	1100	1560
921	649	1200	1660
977	704	1300	1760
1035	762	1400	1860
1089	816	1500	1960

Table A-8. Commercial Wrought Steel Pipe Data (ANSI B36.10)

Nom. Pipe Size	O.D. Inches	Wall Thickness Inches	I.D. Inches
Schedule 10			
1/2	.840	.083	.674
3/4	1.050	.083	.884
1	1.315	.109	1.097
11/2	1.900	.109	1.682
2	2.375	.109	2.157
3	3.500	.120	3.260
4	4.500	.120	4.260
6	6.625	.134	6.357
8	8.625	.148	8.329
10	10.750	.165	10.42
12	12.750	.180	12.39
14	14.00	.250	13.50
16	16.00	.250	15.50
18	18.00	.250	17.50
20	20.00	.250	19.50
24	24.00	.250	23.50
30	30.00	.312	29.38
Schedule 20			
8	8.63	.250	8.13
10	10.80	.250	10.25
12	12.75	.250	12.25
14	14.00	.312	13.38
16	16.00	.312	15.38
18	18.00	.312	17.38
20	20.00	.375	19.25
24	24.00	.375	23.25
30	30.00	.500	29.00
Schedule 40			
1/2	.840	.109	.622
3/4	1.05	.113	.824

Table A-8. Commercial Wrought Steel Pipe Data (ANSI B36.10)

1	1.32	.133	1.05
11/4	1.66	.140	1.38
11/2	1.90	.145	1.61
2	2.38	.154	2.07
1/2	2.88	.203	2.47
3	3.50	.216	3.07
4	4.50	.237	4.03
6	6.63	.280	6.07
8	8.63	.322	7.98
10	10.75	.365	10.02
12	12.75	.406	11.94
14	14.00	.438	13.12
16	16.00	.500	15.00
18	18.00	.562	16.88
20	20.00	.594	18.81
24	24.00	.688	22.62
Schedule 80			
1/2	.840	.147	.546
3/4	1.05	.154	.742
1	1.32	.179	.957
11/4	1.66	.191	1.28
11/2	1.90	.200	1.50
2	2.38	.218	1.94
21/2	2.88	.276	2.32
3	3.50	.300	2.90
4	4.50	.337	3.83
6	6.63	.432	5.76
8	8.63	.500	7.63
10	10.75	.594	9.56
12	12.75	.688	11.37
14	14.00	.750	12.50
16	16.00	.844	14.31
18	18.00	.938	16.12
20	20.00	1.03	17.94
24	24.00	1.22	21.56
Schedule 160			
1/2	.840	.187	.466
3/4	1.05	.219	.614
1	1.32	.250	.815
11/4	1.66	.250	1.16
11/2	1.90	.281	1.34
2	2.38	.344	1.69
21/2	2.88	.375	2.13
3	3.50	.438	2.62

Table A-8. Commercial Wrought Steel Pipe Data (ANSI B36.10)

4	4.50	.531	3.44
6	6.63	.719	5.19
8	8.63	.906	6.81
10	10.75	1.13	8.50
12	12.75	1.31	10.13
14	14.00	1.41	11.19
16	16.00	1.59	12.81
18	18.00	1.78	14.44
20	20.00	1.97	16.06
24	24.00	2.34	19.31

APPENDIX B: CONTROL VALVE STANDARDS AND RECOMMENDED PRACTICES

ISA, 67 ALEXANDER DRIVE, RESEARCH TRIANGLE PARK, NC 27709

ANSI/ISA-75.01.01 (IEC 60534-2-1 Mod)-2007, Flow Equations for Sizing Control Valves

ANSI/ISA-75.02-1996, Control Valve Capacity Test Procedure

ANSI/ISA-75.05.01-2000 (R2005), Control Valve Terminology

ISA-75.07-1997, Laboratory Measurements of Aerodynamic Noise Generated by Control Valves

ANSI/ISA-75.08.01-2002 (R2007) - Face-to-Face Dimensions for Integral Flanged Globe-Style Control Valve Bodies (Classes 125, 150, 250, 300, and 600)

ANSI/ISA-75.08.02-2003 - Face-to-Face Dimensions for Flangeless Control Valves (Classes 150, 300, and 600)

ANSI/ISA-75.08.03-2001 (R2007) - Face-to-Face Dimensions for Socket Weld-End and Screwed-End Globe-Style Control

ANSI/ISA-75.08.04-2001 (R2007) - Face-to-Face Dimensions for Buttweld-End Globe-Style Control Valves (Class 4500)

ANSI/ISA-75.08.06-2002 (R2007) - Face-to-Face Dimensions for Flanged Globe-Style Control Valve Bodies (Classes 900, 1500, and 2500)

ANSI/ISA-75.08.07-2001 (R2007) - Face-to-Face Dimensions for Separable Flanged Globe-Style Control Valves (Classes 150, 300, and 600)

ANSI/ISA-75.08.08-2001 - Face-to-Centerline Dimensions for Flanged Globe-Style Angle Control Valve Bodies (Classes 150, 300, and 600)

ISA-75.10.01-2008, General Requirements for Clamp or Pinch Valves (draft)

ISA-75.10.02-2008. Formerly ANSI/ISA-75.08-1999, Installed Face-to-Face Dimensions for Dual Pinch Flanged Clamp or Pinch Valves

ANSI/ISA-75.11.01-1985 (R2002), Inherent Flow Characteristic and Rangeability of Control Valves (Classes 150, 300, 600, 900, 1500, and 2500)

ANSI/ISA-75.13.01-1996 (R2007) - Method of Evaluating the Performance of Positioners with Analog Input Signals and Pneumatic Output

ISA-75.17-1989 - Control Valve Aerodynamic Noise Prediction

ISA-TR75.04.01-1998 (R2006), Control Valve Position Stability

ANSI/ISA-75.19.01-2007 - Hydrostatic Testing of Control Valves

ISA-RP75.21-1989 - (R1996) - Process Data Presentation for Control Valves

FLUID CONTROLS INSTITUTE (FCI), P.O. 1300 SUMNER AVENUE, CLEVELAND, OHIO 44115.

ANSI/FCI 70-2-2006, Control Valve Seat Leakage

FCI 84-1-1985, Metric Definition of the Valve Flow Coefficient C_v

FCI 87-2-1990 (R1998), Power Signal Standard for Spring-Diaphragm Actuated Control Valves

FCI 91-1-1997 (R2003), Standard for Qualification of Control Valve Stem Seals

INTERNATIONAL ELECTROTECHNICAL COMMISSION (IEC), GENEVA, SWITZERLAND

Industrial Process Control Valves:

IEC 60534-1, Part 1: Control Valve Terminology and General Considerations

IEC 60534-2-1, Part 2: Flow Capacity. Section One: Sizing Equations for Fluid Flow Under Installed Conditions

IEC 60534-2-3, Part 2: Flow Capacity. Section Three: Test Procedures

IEC 60534-2-4, Part 2: Flow Capacity. Section Four: Inherent Flow Characteristics and Rangeability

IEC 60534-3-1, Part 3: Dimensions. Section One: Face-to-Face Dimensions for Flanged, Two-Way, Globe-Type Straight Pattern and Centre-to-Face Dimensions for Flanged, Two-Way, Globe-Type, Angle Pattern Control Valves

IEC 60534-3-2, Part 3: Dimensions. Section Two: Face-to-Face Dimensions for Rotary Control Valves Except Butterfly Valves

IEC 60534-4, Part 4: Inspection and Routine Testing

IEC 60534-5, Part 5: Marking

IEC 60534-6-1, Part 6: Mounting Details for Attachment of Positioners to Control Valves – Section 1: Positioner Mounting on Linear Actuators

IEC 60534-7, Part 7: Control Valve Data Sheet

IEC 60534-8-1, Part 8: Noise Considerations. Section One: Laboratory Measurement of Noise Generated by Aerodynamic Flow Through Control Valves

IEC 60534-8-2, Part 8: Noise Considerations. Section Two: Laboratory Measurement of Noise Generated by Hydrodynamic Flow through Control Valves

IEC 60534-8-3, Part 8: Noise Considerations. Section Three: Control Valve Aerodynamic Noise Prediction Method

IEC 60534-8-4, Part 8: Noise Considerations. Section Four: Prediction of Noise Generated by Hydrodynamic Flow

MANUFACTURERS STANDARDIZATION SOCIETY OF THE VALVE AND FITTINGS INDUSTRY (MSS), 127 PARK ST., N.E., VIENNA, VA 22180

SP-25-1998, Standard Marking Systems for Valves, Fittings, Flanges, and Unions

SP-53-1999 (R2007), Quality Standard for Steel Castings and Forgings for Valves, Flanges and Fittings and Other Piping Components - Magnetic Particle Exam Method

SP-54-1999 (R2007), Quality Standard for Steel Castings for Valves, Flanges, and Fittings and Other Piping Components - Radiographic Examination Method

THE AMERICAN SOCIETY OF MECHANICAL ENGINEERS (ASME/ANSI), 3 PARK AVENUE, NEW YORK, NY 10016-5990

B16.5-2003, Pipe Flanges and Flanged Fittings

B16.34-2004, Valves - Flanged, Threaded, and Welding End

B31.1-2007, Power Piping

BPE-2007, Bioprocessing Equipment

AMERICAN SOCIETY FOR TESTING AND MATERIALS (ASTM) INTERNATIONAL, 100 BARR HARBOR DRIVE, PO BOX C700, WEST CONSHOHOCKEN, PA, 19428-2959

Standard Specification for:

A105, Carbon Steel Forgings for Piping Applications

A182, Forged or Rolled Alloy and Stainless Steel Pipe Flanges, Forged Fittings, and Valves and Parts for High-Temperature Service

A193, Alloy-Steel and Stainless Steel Bolting Materials for High Temperature or High Pressure Service and Other Special Purpose Applications

A194, Carbon and Alloy Steel Nuts for Bolts for High-Pressure or High-Temperature Service, or Both

A216, Steel Castings, Carbon, Suitable for Fusion Welding, for High-Temperature Service

A217, Steel Castings, Martensitic Stainless and Alloy, for Pressure-Containing Parts, Suitable for High-Temperature Service

INDEX